T0186320

A Guide for the
Design and Management of
Combined Sewerage Networks

State of the Art

A Guide for the
Design and Management of
Combined Sewerage Networks
State of the Art

François Valiron and Michel Affholder

A.A. BALKEMA/ROTTERDAM/BROOKFIELD/2001

Ouvrage publié avec le concours du Ministère français
chargè de la Culture - Centre national du livre

Translation of: *Guide de conception et de gestion Des Réseaux d'assainissement unitaires*
état de l'art
©Technique & Documentation, 1996 Paris.

Authorization to photocopy items for internal or personal use, or the internal or personal use of specific clients, is granted by A.A. Balkema, Rotterdam, provided that the base fee of US$1.50 per copy, plus US$0.10 per page is paid directly to Copyright Clearance Center, 222 Rosewood Drive, Danvers, MA 01923, USA. For those organizations that have been granted a photocopy licence by CCC, a separate system of payment has been arranged. The fee code for users of the Transactional Reporting Service is: 90 5809 225 9/2001 US$1.50 + US$0.10.

A.A. Balkema, P.O. Box 1675, 3000 BR Rotterdam, Netherlands
Fax: +31.10.4135947; E-mail: balkema@balkema.nl

Distributed in USA and Canada by
A.A. Balkema Publishers, 2252 Ridge Road, Brookfield, Vermont 05036 USA.
Fax: 802.276.3837; E-mail: Info@ashgate.com

ISBN 90 5809 225 9

© 2001 Copyright Reserved

PREFACE

L'Agence de l'Eau Seine-Normandie (Water Bureau of the Seine River, Normandy) was motivated by the following three considerations to undertake an in-depth study of storm weirs and management of combined sewerage networks.

— **Complexity of management of combined sewerage networks** during the rainy season (75% of the total area) and the impact of storm overflows on the quality of the river. For example, rainfall on 26 and 27 June 1990 in the region of Ile de France added to the natural environment twice the amount of pollution produced in one day during the dry season.

— **The decree of 23 March 1993** stipulated that an authorisation would be required for overflow discharges of storm weirs in regions with an estimated population of more than 2000 by 1997 and a declaration and provision of facilities for measurements would be required in other cases.

— **The new law regarding water, formulated in January 1992**, which provides an incentive for overall management of networks and waste water treatment plants.

Undertaking this in-depth study was prompted by the British experience in management of overflow weirs and a survey conducted in Ile de France by **M. Valiron** in management of weirs in five districts (SIAAP and districts 92, 93, 94 and 75).

During the same period, around 1995, the Group at AGHTM was working on similar topics. A good liaison existed between the two groups because **M. Tabuchi**, who succeeded **M. Gauvent**, also directed the work carried out at the Bureau.

Thus AGHTM and the Bureau decided to meet on 19 May 1995 at Achères to exchange information on the management of networks and weirs in the Parisian region. Nine case studies were presented, covering the British survey of the Parisian region as well as a technical guide.

In this meeting the discussion centred on **the crucial role of storm weirs**, the improvements needed for reducing pollution and the importance of such weirs in the **management of combined sewerage networks**. The discussion also covered the importance of different points of exchange of flows, sometimes quite

v

mystifying, in the main sewers of the network, which were historically considered very complex. Reduction of wastes added during the rainy season to the natural environment requires a sound knowledge of these works since there are numerous possibilities for increasing their effectiveness and improving their management.

For these reasons, it was considered useful to present a compilation of these studies and the discussion in the form of a **Guide for the Design and Management of Combined Sewerage Networks**. Hopefully it will find a place among the collection of papers on control of urban pollution during the rainy season inaugurated in December 1992, entitled the **State of the Art** by **M. Valiron** and **M. Tabuchi**, published under the double patronage of the Water Bureau and the General Association of Public Health Scientists and Engineers.

P. F. Téniere-Buchot
Director,
Water Bureau of Seine, Normandy

M. Affholder
President,
Association of Public Health
Scientists and Engineers

CONTENTS

INTRODUCTION

This Guide deals with structures which during the rainy season discharge into a combined sewerage network which cannot be carried farther downstream without causing flooding.

When the wastes are discharged into the natural environment, the structures function as **storm weirs**; if the wastes are discharged into other sewers, they function as **exchange works**, called **'bypasses'**, which serve as a mesh in the network. By extension, devices of this type may be used for collecting water during the dry season only into a combined sewer and later diverting it to a sanitary sewer. This topic is also included in the present discussion. Obviously, all complementary works, such as pumping stations, which can improve the functioning of these 'weirs' in a broad sense or may assist in discharging the function of combating overflow of the network in case of flooding are also included.

These works may also be designed for controlling polluting wastes, thanks to their characteristic architecture or complementary works such as settling chambers, screens, basins of storage/decantation etc. **This additional function of removal of pollutants has been paid due attention only recently because the impact of wastes during the rainy season had long been overshadowed by the effects of untreated or incompletely treated waste water from other sources.**

Today it is recognised throughout the world that the wastes of numerous improperly operated weirs are responsible for deterioration of water in rivers compared to what could be expected in view of the treatment of waste water from other sources, not only during the rainy season but also during the dry season, to preclude the long-term effects of pollution brought in by rainstorms.

Thus weirs and associated works play a crucial role in the effectiveness and management of the network vis-à-vis the protection of the natural environment.

The control of these weirs is complex and requires time and large financial input. In England where this problem received attention about 20 years ago, 8000 of the 25,000 weirs of all types are today still defective and in need of improvement or modification. In France, where neither the total number of existing weirs nor their impact is properly known, it would seem that a large number of these works, if not almost all, need to be tested and modernised; many need to be provided with a storage/decantation basin. Among the thousand or so existing in the four districts covered by the survey conducted in Ile de France, less than 50 are equipped with adjustable valves and very few are provided with facilities for measurements.

Application of the decree of 23 March 1993 will oblige the managers to take inventory of their weirs and to equip many of them with facilities for measurements; those which need to be modernised have to be identified so that spread of wastes and pollution is reduced. Finally, the renovated weirs should be taken into account in the overall management of the system.

This Guide, based on the experience acquired in Ile de France and the earlier experience of the British as well as the up-to-date knowledge acquired from more than a hundred French and foreign publications, will hopefully be useful to French managers in considering the new obligatory legal regulations.

The book comprises seven chapters and a Bibliography listing recent documents cited in the text; it will be immensely helpful to the reader in deciding which improvements need to be carried out in the weirs under his charge and managing the network.

1. **Some data on the pollutions generated** in nature and their impact in the case of combined and rainstorm networks.
2. **Different types of weirs and their accessories and complementary works:** settling chambers, skimmers and screens, valves, sewers and storage/decantation basins.
3. **Computation and measurements of weirs**: computational methods and their limitations, measurement of liquid flux and pollution with the help of standard devices, furnishing all data necessary for their management, including data on pollution.
4. **Evaluation and improvement of hydraulic and remediation efficiency.** State of knowledge and improvements possible, methodology, choice of device, equipment for valves, storm basins etc.
5. **Location and calibration of weirs**, choice of a work and improvement. Role of modelling of the network.
6. **Legal regulations concerning wastes** in France and England; contents of the French decrees of March 1993. Problems to be resolved.
7. **Conclusions and strategy proposed** for improved knowledge and compliance with the regulations.

Finally, two additional chapters:

— **Annexure I contains computational tables and methods for dimensioning storm basins.**
— **Annexure II reproduces the nine case studies or surveys presented during the one-day meeting at Achères,** which elaborate certain points discussed in the seven chapters of the Guide:

Case 1: Present state of knowledge on the impact of wastes in Seine, a follow-up of the work of PIREN Seine.

Case 2: Modernisation of Parisian weirs.

Case 3: Instructions for the programme of measurement carried out in 1993 on the Parisian weirs.

Case 4: Measurements of pollution carried out in Val de Marne, especially in the basin of Vitry.

Case 5: Summary of surveys of weirs in Ile de France.

Case 6: Management of weirs in the Boulogne loop.

Case 7: Improvement of interdistrict works of SIAAP.

Case 8: The new directive plan of Achères.

Case 9: Plan for overall strategy of sanitation in Seine Saint Denis.

The Guide includes references for each case as well as an index (of key words), which will help the reader to acquire in-depth knowledge if he so desires.

This book was designed and edited under the direction of Michel Affholder (Paris) and François Valiron (Honorary professor at the Ecole National des Ponts et Chaussées).

The Guide for design and management of a combined sanitary network as well as Annexure I were edited by F. Valiron while Annexure II was edited by the following team:

Mme C. Cogez (DEA of Seine Saint Denis)

Melle Genestine (SIAAP)

Messrs.	F. Bethouart (Paris City)
	B. Breuil (DEA of Seine Saint Denis)
	A. Constat (Paris City)
	M. Gousailles (SIAAP)
	M. Le Scoarnec (SIAAP)
	J. M. Mouchel (Cergrene)
	A. Rabier (DSEA Val de Marne)
	M. Soulier (DEA of Hauts de Seine)
	F. Valiron
	M. Van Den Boght (Aquarius)
	M. Wollart (Aquarius)

CHAPTER 1

Some Data on Rainstorm Run-off Pollution

1. SOURCES OF POLLUTION FROM RAINSTORM RUN-OFF
   (see Valiron, ref. A)

Pollution from rainstorm run-off comprises in part the **atmospheric pollution** dissolved in rain and the effects of rainfall on land during its **overland flow**.

1.1 ATMOSPHERIC POLLUTION AND RAIN

Atmospheric pollution originates from emissions of industry and heating systems and the volatile part of the exhausts from engines and combustion processes. In France it was estimated that of the 16.5 million tons of polluting emissions generated in 1983-1984, one-third was contributed by the transport system. These emissions, containing about 35% hydrocarbons and nitrogen oxide and 30% sulphur dioxide, spread in the atmosphere; rain dissolves a part of them, the quantity varying with the season and location (but there is no close correlation with the site of emission). Table 1.1 shows the range of variation of major contaminants measured in precipitations.

In the pollutants in urban wastes, during the rainy season the contribution of the contaminants listed in Table 1.1 is, on average, 20/25% while that of the heavy metals may be as high as 70/75%.

pH	4 to 7
COD	20 to 30 mg O_2/l
SO_4	2 to 35 mg/l
Cu	0.5 to 2 mg/l
Na	0.5 to 2 mg/l
Zn	0.008 to 0.02 mg/l
Pb	0 to 0.15 mg/l

Table 1.1

The other part of rainstorm pollution comes from rain-water runoff on land and vegetation; the water may wash—or even erode—the soil. In this manner, various wastes generated by human activity, including the heavy dust carried in smoke and the solid particles in vehicular exhaust, are removed from the soil.

The quantity of these contaminants depends strongly on the site; in a town replete with enterprises and large impermeable surfaces, the concentration of pollution per hectare is much higher than in rural areas.

Table 1.2, based on American data (**Desbordes**, ref. 1) shows the order of magnitude of pollution accumulated on urban roads, expressed as ppm (except the solids, expressed in kg/km²/yr and faecal coliforms, expressed in number/ 100 ml).

	Residential low density individual	Residential high density collective	Commercial	Small industry	Roads and roadways
Solid matter (kg/km²/yr)	10-180	30-210	13-180	80-290	13-1100
BOD$_5$ (ppm)	5200	3300	7100	2900	2300-10,000
COD (ppm)	40,000	40/42,000	39/62,000	25,000	53/80,000
Total N (ppm)	480	55/600	400	430	220/1000
Pb (ppm)	1570	1900	2300	1600	450/2300
Cd (ppm)	3.2	2.7	2.9	3.6	2.1/10.2
Faecal coliforms (number/100 ml)	60/82,000	25/32,000	36,000	30,000	19/38,000

Table 1.2: Pollution accumulated on urban roads

The impact of the contribution of roadways on rainstorm pollution is very significant, between 40 and 50% of the total pollutants in the run-off; in residential areas it is about 30% and for other users of the land, such as industry and large commercial complexes, more than 90%.

The concentration of pollutants in rain-water run-off and the pollutional load vary with the amount of rainfall and the time elapsed since the last storm (the deposits to be removed being roughly proportional to the duration of the dry period preceding the rain).

Table 1.3 shows the mean concentrations and their variance of the median event of rain-water flow by zone; the data were taken from studies conducted in 22 towns of the USA in 1981-1982 under the National Urban Run-off Programme (NURP).

Pollutant	Mean	Var.	Mean	Var.	Mean	Var.	Mean	Var.
BOD mg/l	10,000	0.41	7800	0.52	9300	0.31	—	—
COD mg/l	73,000	0.55	65,000	0.58	57,000	0.39	40,000	0.78
SS mg/l	101,100	0.96	67,000	1.10	69,000	0.85	70,000	2.90
Pb mg/l	144	0.75	114	1.40	104	0.68	34	1.50
Cu mg/l	33	0.99	27	1.30	29	0.81	—	—
Zn mg/l	135	0.84	154	0.78	226	1.10	195	0.66
TKN mg/l	1900	0.73	1290	0.50	1180	0.43	965	1.00
NO$_{2+3}$ mg/l	736	0.83	558	0.67	572	0.48	543	0.91
Total P mg/l	383	0.69	263	0.75	201	0.57	121	1.70
Particulate P mg/l	143	0.46	56	0.75	80	0.71	26	2.10

Table 1.3: *Mean concentrations in rain-water flow*

Table 1.4, extracted from the results of measurements carried out at 6 sites in France, shows large variations from one site to another in terms of mean concentrations and annual amount of rainstorm run-off. Observations made in the USA are fully confirmed by these data.

Finally, on the same site, **contamination carried in run-off from a single rainfall may represent 20 to 25 %**, even 50%, of the **annual amount**. The mean concentrations of an episode may finally be 5 to 10 times the mean annual concentration. This highlights the difficulties associated with measurements which need to be carried out over a long period and the random character of rainstorm run-off pollution.

2. RAINSTORM RUN-OFF POLLUTION IN THE NETWORK

2.1 SUSPENDED MATTER, DEPOSITED MATTER AND RESUSPENSION OF MATTER

• **Suspended matter**

The size of the suspended matter and its velocity of fall explain that a significant portion of it gets deposited (Fig. 1.1).

Basin		Contaminant				
		SS	COD	BOD_s	TKN	Pb
Maurepas	C	190	77	12	3.3	0.085
	M	940	380	55	16	0.41
Les Ulis Nord	C	440	190	34	6.1	0.12
	M	1100	460	85	17	0.30
Aix Zup	C	300	200	38	5.4	0.16
	M	630	430	75	12	0.35
Vélizy	C	190	90	17	3.8	0.47
	M	400	190	36	8.0	1.0
Nice	C	130	120	28	—	—
Barron de Berre	M	540	530	—	—	—
Admissible concentration		30(a)	120(a)	40(a)	15(b)	0.05(c)

(a) Norm of waste of treatment plant—level E (circular of 4.11.80).
(b) Norm of waste of treatment plant—level 2 (circular of 29.11.80).
(c) Quality required of surficial water used for production of potable water (CEE directive of 16.6.76).

Table 1.4: *Mean concentration (in mg/l) and specific annual amount (in kg/ha/yr) of rainstorm run-off*

Depending on the storm and whether the network is separate or combined, 70 to 80% of particles are smaller than 40 µm (according to **Stahre**, ref. K) or 100 µm (according to **Chebbo**, ref. 3). The velocity of fall increases with the concentration and hence the time of decantation becomes shorter (5 h for a concentration of 100 mg/l and 1.5 h for 215 mg/l).

Finally, it appears that certain elements of rainstorm run-off play the role of agent of flocculation, helping the fine particles to agglomerate; hence an increase in the size resulting in acceleration of their fall.

Another fact observed by many researchers is that a large part of the pollution is associated with SS, with the exception of nitrates and soluble phosphorus; hence an entrainment of 80 to 90% of COD, 65 to 80% of TKN, 80 to 95% of hydrocarbons and a near totality of lead.

According to **Chebbo** (ref. 4), the deposited material is more than 20% of the mass of yearly wastes, except for HC, as indicated in Table 1.5.

According to **Bachoc** (ref. 4), for an average rainfall the deposited matter seems to provide 30 to 45% of the total solids in suspension and 40 to 42% of

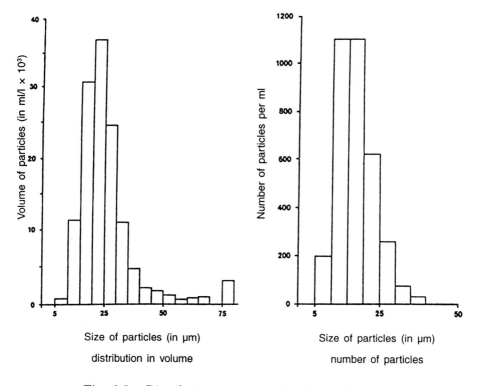

Fig. 1.1: *Distribution in size and volume of suspended matter in a storm run-off.*

24%	SS
22%	COD or BOD$_5$
14%	HC
28%	lead

Table 1.5

volatile matter. It may also be noted that the cohesion of the deposited matter increases strongly with temperature; hence increased velocity of circulation results in its conversion to suspended matter.

• **Resuspension of deposited matter**

Resuspension of deposited matter and its transport by saltation thus take place primarily at points of the network constructed for favouring the deposits, thanks to a settling zone created by a constriction; resuspension occurs when the velocity is high compared to the flow rate before the operation.

Recovery of the deposited matter by saltation occurs at two particular instants: when the level of contaminants is low and at the start of the next rainfall. **It will be seen in Chapters 3 and 4 that this explains the better efficiency of high weirs.**

2.2 VARIATION IN WATER CONTAMINATION AND POLLUTION DURING A STORM EPISODE

During rainfall it is generally noted that at some point in the network a divergence occurs between the variation of concentration and that of the flow rate (Fig. 1.2a).

From these two curves, another curve can be drawn which expresses the ratio of contaminated mass to total mass for the episode as a function of the ratio of the volume that flows out to total volume of the event (Fig. 1.2b).

An M(V) curve may be drawn for each rainfall and such curves may be superposed (Fig. 1.3); this procedure may be repeated for various contaminants (Fig. 1.4).

It can be seen from Fig. 1.4 that the M(V) curves for the same event differ for different contaminants.

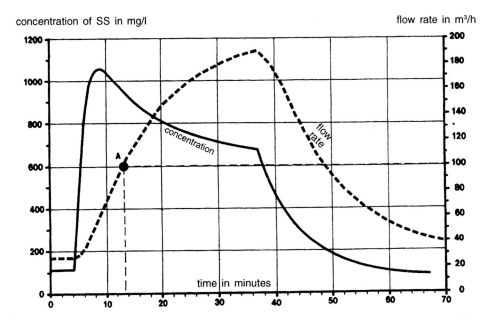

Fig. 1.2a: Hydrograph and pollutograph.
Mantes la Ville—Rainfall of 17.10.1976

6

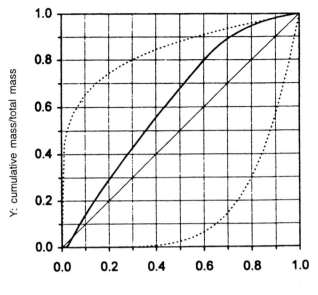

X: Cumulative volume/total volume

Fig. 1.2b: *The M(V) curve.*
Mantes la Ville—Rainfall of 17.10.1976

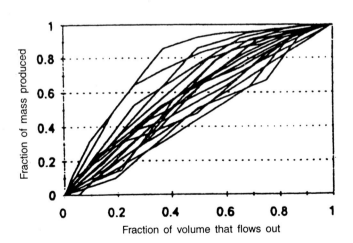

Fraction of volume that flows out

Fig. 1.3: *Superposition of M(V) curves for SS for data from Vélizy*
(Saget and Chebbo, ref. 5).

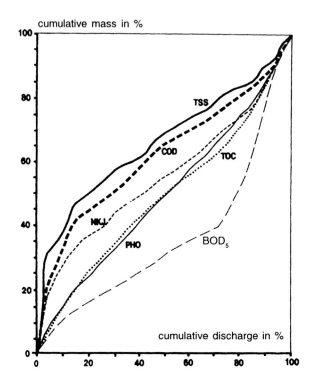

Fig. 1.4: M(V) curves for various contaminants (Geiger, ref. 7)

An M(V) curve may be drawn using values averaged over the events. Fig. 1.5 shows that the pattern of these curves differs from one site to another (and varies with the point of the network at which the measurements are taken). **The positive divergence** of these curves from the equilibrium curve shows that for a flow representing, say 40% of the total volume, the contaminating mass varies from 46 to 74%.

Such a sharp increase in polluting flux compared to the flow rate is not a general phenomenon. It can be seen from Fig. 1.6 that in some cases the reverse may take place.

The sharp increase in polluting mass seems to occur more often in the case of small basins (smaller than 10 ha); this phenomenon is sometimes termed the **'first tide'**—a rather vague description. But it is logical since it occurs systematically if the measurement is carried out close to the flow stream. According to some researchers this phenomenon occurs more frequently in steeply sloping zones.

Thus for an efficient management one should use the basic data corresponding to several storm episodes in order to compile hydrographs Q(t), pollutographs

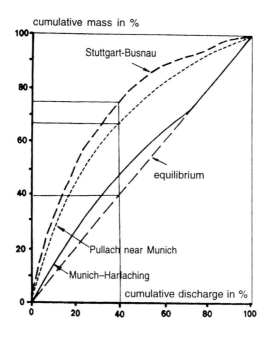

Fig. 1.5: *Comparison of M(V) curves for different sites (Geiger, ref. 7).*

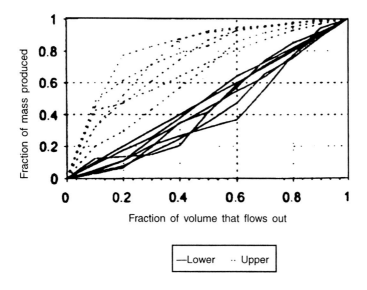

Fig. 1.6: *Lower and upper M(V) curves for SS corresponding to 6 sites (Saget and Chebbo, 1994).*

C(t) and hyetographs P(t) which show the rainfall and the preceding dry period. This makes it possible to draw M(V) curves (see above) as well as other curves which are useful in management of weirs. Thus a solid base is provided for the vague concept of 'first tide', indeed a rare phenomenon.

As a matter of fact, knowledge of these different curves is useful in the management of a weir situated at the point of measurement and computation of the volume of the eventual 'storage/decantation basin' associated with the weir.

For further details, the reader may see **Bertrand-Krajewski** (ref. 5).

It may be noted that the M(V) curves for separate and combined networks are similar.

Computation of the volume of storage required to intercept x% of the mass of a given pollutant is determined from the corresponding M(V) curve.

Depending on whether one uses the curve of averages over the year or that averaged over a storm episode of a certain frequency, one obtains the volume of storage corresponding to the chosen period.

For example, **Chebbo** (ref. 4) gives the volume to be stored for intercepting 80% of the annual mass of SS in the form given in Table 1.6 (expressed in m^3/ha impermeable area).

Type of network	Minimum	Average	Maximum
Combined	33	51	64

Table 1.6: Volume to be stored for intercepting 80% of annual SS (in m^3/ha impermeable area)

• **In the case of a weir** (Fig. 1.7) it begins to overflow when the flow rate exceeds Q_d during a time interval $t_2 - t_1$ with flow rate varying between Q_d and Q_{max} (curve 2); during this overflow the concentration of contaminants varies from C_1 to C_2 (curve 1). The mass of SS carried to a location before t_1 and after t_2 is given in percentage as $[100—(M_2—M_1)]$ (curve 3) and the volume in percentage as $[100—(V_2—V_1)]$ (curve 4).

During the overflow the weir sends volume $Q_d \times (t_2 - t_1)$ to the plant and the hatched volume to the natural environment.

Assuming that the weir divides the polluting flux in proportion to the volume, the polluting mass sent to the natural environment is given by:

$$\text{Mass rejected during the storm} = \int_{t_1}^{t_2} [Q - Q_d] \, C(t)dt$$

Thus its value can be computed from curves 1 to 4.

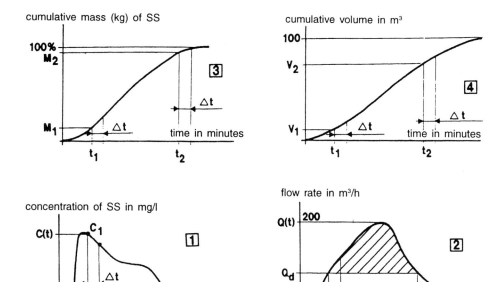

Fig. 1.7: *C(t), O(t), M(t) and U(t) curves of a storm event.*

A higher weir begins to overflow at much higher Q, thanks to the time Δt spent in filling up the additional storage trench ΔS related approximately to the height of the weir by the formula

$$\Delta t = \frac{\Delta S}{Q_d}$$

Curve 2 of Fig. 1.7 shows that when the level drops, there is a delay in t; the interval $t_2 - t_1$ remains the same while the mean concentration is lower (curve 1). Curves 3 and 4 also show that the quantity sent to the plant increases and thus the amount spread in the natural environment is reduced.

3.   WASTES FROM RAIN-WATER OVERFLOW

Table 1.7, taken from **Cottet** (ref. 8), compares the annual, daily and hourly contaminants in a rainstorm run-off with the amount of waste water in the same region; the data refer to observations made at Ulis and Maurepas.

The table shows that the average yearly contribution of rainstorm run-off is much smaller than that of waste water with reference to SS, BOD_5, COD, TKN and equal to or even larger in terms of heavy metals; the contribution of storms of small duration is much larger, however.

11

	Annual base	**Daily base**	**Hourly base**
SS	ER = EU/2	ER = EU	ER = 50 × EU
BOD$_5$	ER = EU/27	ER = EU/6	ER = 4 × EU
COD	ER = EU/9	ER = EU/2	ER = 12 × EU
TKN	ER = EU/27	ER = EU/7	ER = 3.5 × EU
Heavy metals			
Pb	ER = 27 × EU	ER = 80 × EU	ER = 2000 × EU
Zn	ER = EU	ER = 4 × EU	ER = 100 × EU
Cu	ER = EU/4.5	ER = EU/2	ER = 15 × EU
Cr	ER = EU/4	ER = EU/1.5	ER = 16 × EU
Hg	ER = EU	ER = 7 × EU	
Cds	ER = EU	ER = 5 × EU	

EU : amount in waste water
ER : amount in run-off

Table 1.7: *Comparison of quantity of contaminants in waste water versus run-off.*

The values shown in Table 1.7 are only indicative because, as explained in Sections 1 and 2, they vary from one site to another and also depend on whether the network is combined or separate.

• An evaluation of wastes from weirs as communicated recently by Tabuchi (ref. 9) for a town of 100,000 inhabitants is given below.

The pollution carried by the storm weirs to the natural environment needs to be defined in quality as well as quantity. Today it may be considered that the characteristics of pollution are known and all that is required is to consult the bibliographical data on the quality of overflow of combined networks. Nevertheless, it is worthwhile to summarise the important characteristics: a composition close to that of waste water of the dry season with some peculiarities, however, such as a sufficiently large variation in high concentration of suspended matter, a lower concentration of TKN and reduced conductivity. Furthermore, the composition varies with the duration of precipitation, tending to match the composition of the rainstorm. Along with this physicochemical polllution, one should also consider the visual pollution carried by the weir overflow in the form of floating garbage (plastic bottles, paper, wood, fats, polystyrene etc.). Of course, this pollution is not very damaging in terms of impact on the ecosystem but it tarnishes significantly the image of sanitation.

Evaluation of the quantity of pollution discharged by a community to the natural environment is a rather difficult task because it depends strongly on the characteristics of the sanitation system and the relevant catchment area. Such

12

evaluation is necessary, however, for several reasons: dimensioning works and assessing the impact on the receiving environment and, also, the new regulations oblige prefects on the one hand to specify the objectives of treatment and, on the other, the community to carry out diagnostic studies of the network for determining the polluting flux brought into the natural environment. The first step for such evaluation is to determine ratios of pollutant load to rainfall but the results must subsequently be confirmed by measurement and, if possible, by mathematical simulations.

The measurements carried out by SIAAP on the weir of Clichy from 1990 to 1991 enabled computation of the ratios of pollution that may be considered representative of the mass of pollution spread by a combined network, taking into account the size of the catchment area; the results obtained are presented in Table 1.8.

SS	BOD$_s$	COD	TKN	Pt	Cd	Cu	Pb
1.20	0.36	1.30	0.075	0.018	1.2×10^{-5}	7.1×10^{-4}	1.1×10^{-3}

Table 1.8: Results of measurements carried out at Clichy (1990-1991)
(Flux expressed in kg/mm rainfall/impermeable ha)

Using these ratios, the theoretical case of a town with 100,000 inhabitants was studied. The impermeable catchment area was computed on the basis of 100 m^2 per person. Of the 600 mm of total annual rainfall, 400 mm resulted in overflow. The storm weir was so dimensioned as to completely preclude the spread of wastes to the natural environment and for this the discharge had to be kept smaller than twice the mean discharge during the dry season.

The annual flux of pollution was evaluated with the help of averages computed per mm of rainfall per impermeable hectare area, and expressed on the basis of 400 mm rainfall and 1000 impermeable hectare area.

The annual pollution of the rainy season is compared with the annual pollution during the dry season from a treatment plant whose wastes conform to the stipulations of the European directive.

A comparison of the polluting fluxes released as a function of the amount of rainfall was also done.

The results obtained are compared with the pollution released daily in the natural environment during the dry season. All the data are presented in Table 1.9.

	Pollution during the rainy season in kg/yr	Pollution during the dry season in kg/yr
Substance	*Overflow*	*Released STEP*
SS	470,000	320,000
BOD$_5$	144,000	230,000
COD	520,000	1,140,000
TKN	30,000	91,000
Total P	7000	18,000
Cd	5	2
Cu	280	137
Pb	460	80

Table 1.9: *Comparison of annual pollutions with those of STEP*

• **An actual case for which evaluation and some accurate figures are presently available is that of contamination reaching the plant of Achères in the Parisian region.**

This important plant, described in case study 7 by **M. Goussailles**, is capable of treating 2,100,000 m³/day during the dry season and is fed by 5 large sewers which in 1993 carried:

905 million m³ in the entire period,

845 million m³ during the dry period.

Thus the volume carried during the rainy period was 60 million m³. The average flow rate during the rainy period is also accurately known (34.9 m³/s); hence one can deduce the cumulative duration of the rainy period in days (19.8) corresponding to more than 50 storm episodes (see Table 1.10).

It is estimated that of 900 km² of the catchment area, 500 km² are impermeable. Further, of these 500 km², 170 km² is under a separate storm-water system and 330 km² under a combined system. Also, there is a rain-water flow equivalent to 427 mm (610 mm rainfall with run-off coefficient equal to 0.7). Thus there is a flow of 72 million m³ into the natural environment from rainstorms and 80 million m³ via the weirs excluding that at Achères (which accounts for 140 million m³ of rain-water flow, less the 60 million m³ diverted to Achères).

The annual polluting flux **from the weirs alone** is of the order of 24,000 tons (SS, COD, TKN), say **an average of 480 tons per rainfall,** while the amount of effluents treated at Achères is around 54,000 tons per annum or **147 tons per day. The inflow during the rainy period is almost double if one takes into account the storm sewers** carrying 45,000 tons or, on average, **900 tons per event; this is 6 times the discharge at Achères.**

14

Substance	Pollution added during the rainy period in kg/day			Pollution of dry period in kg/day	
	5 mm rainfall	*10 mm rainfall*	*20 mm rainfall*	*Rough figure*	*Released STEP*
SS	5800	11,700	23,400	9000	875
BOD$_5$	1800	3600	7220	7000	625
COD	6500	13,000	26,000	17,000	3125
TKN	370	740	1490	1500	250
Total P	90	180	370	400	50
Cd	0.06	0.12	0.24	0.05	0.01
Cu	4	7	14	2.9	0.38
Pb	6	11	23	1.7	0.21

Table 1.10: Comparison of different amounts of rainfall

• **Another actual case is that of the Urban Community of Lyon. Hodeaux** (ref. 10) estimated the discharge during the rainy period and the contaminants released during the rainy period and compared with those released during the dry period (by inhabitants and industry) in Table 1.11.

	SS	**BOD$_s$**	**COD**	**Pb**	**Hgd**	**NjK**
Discharge during rainy period	30,000	5000	20,000	40	600	1000
Discharge during dry period	15,000	20,000	45,000	0.3	—	10,000
Ratio of discharge: rainy period/dry period	2	0.4	0.44	130	—	0.1

Table 1.11: Annual average discharge

The volume released during the rainy period is the same as during the dry period (120 million m^3); this is explained by a much lower density of land occupancy compared to Achères.

• **The measurements made at Metz** for calibrating the model of the network (**Safège**, ref. 4.3) showed that for the 20 storm events, representing 20% of the rainfall and one-sixth of the number of storm events in a year, the pollution released is equivalent to 93,000 inhabitants. The annual waste is equivalent to that from the treatment plant.

4. IMPACT OF RUN-OFF WASTES ON THE NATURAL ENVIRONMENT

The components of rainstorm run-off are essentially the same as those of waste water, though the proportions differ in the two cases. The overflow from combined systems shows greater similarity. It is therefore understandable that the general laws governing the impacts of these waters on the natural environment are identical.

Table 1.12 gives a schematic indication of the zones of the river downstream of the point of release where the effects of its various components are produced.

Local 10 m to 2000 or 3000 m	Hydraulic flux Flotsam, bacteria, SS
Regional 2 or 3 km to 50 or 100 km	SS, bacteria, dissolved oxygen, nutrients, dissolved toxic salts
Basin and beyond	Dissolved toxic salts, nutrients

Table 1.12: Spatial impacts of wastes

Urban wastes during the rainy period differ from those of dry weather primarily with respect to their periodic nature and the much larger concentration of certain substances, especially SS and heavy metals.

• **Periodic nature** implies effects of shocks leading to transient but definitive pollutants as well as resuspension of deposited material during the release of waste water, followed by sedimentation when the overflow ceases. This results in increased importance of the effect of the phenomenon of exchange between water and the material deposited on the bottom of the waterbody.

• **High concentration** of SS, especially of toxic heavy metals, has a long-term effect on fauna due to accumulation in living matter.

Figure 1.8, taken from **Ellis** (ref. 11), distinguishes between the dissolved form and particulate form of heavy metals released and highlights the dangers of these wastes, particularly in the urban zone. Their concentration almost always exceeds the limits prescribed for Zn, Cd and Cu before dilution in the receiving environment.

These differences in waste water arise primarily from the **periodic nature** of the run-off wastes, which cause transient shocks, followed by a long-term effect: the shorter the duration of exposure, the lower the concentration of dissolved oxygen tolerated by fauna. These differences result in **immediate** or short-term as well as delayed or **long-term effects.** The effects of rainstorm run-off on the natural environment depend on three parameters—intensity, duration and frequency—while for waste water, intensity is the only parameter of consequence.

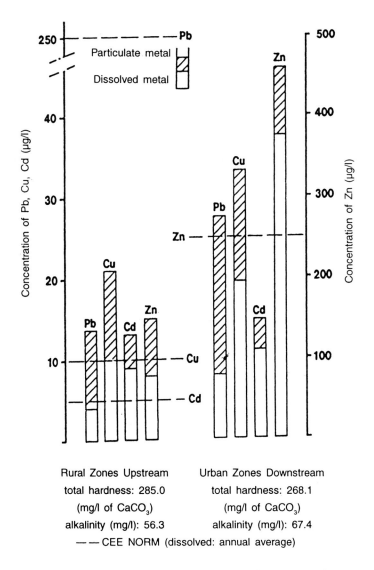

Rural Zones Upstream
total hardness: 285.0
(mg/l of CaCO₃)
alkalinity (mg/l): 56.3

Urban Zones Downstream
total hardness: 268.1
(mg/l of CaCO₃)
alkalinity (mg/l): 67.4

— — CEE NORM (dissolved: annual average)

Fig. 1.8: Comparison of mean concentration of heavy metals in rainstorm run-off according to the CEE norm for water potability.

17

These effects cease as soon as the source is removed; there is no residual effect of long-term duration.

It may be noted that the above holds true for accidental pollutions also. **The importance of the impact of this pollution should be assessed from the statistically largest flux.**

It seems that the most spectacular impact is seen in fish mortality due to asphyxiation resulting from a shortage of oxygen. Decomposition of organic compounds of carbon and nitrogen requires oxygen and thereby a shortage is created for fish.

As shown in Table 1.13, the reaction of fish to low concentration of O_2 depends on the species, duration of exposure, water temperature and physiological state of the animal. The effects are more intense in the early stages of growth. It has further been reported that repeated exposure to low concentrations of O_2 has a detrimental effect on growth and reproduction.

Species	Duration of exposure	Temperature °C		
		10°	*16°*	*20°*
Rainbow trout	3.5 hours	1.7 (1.2)	1.9 (1.5)	2.1 (1.6)
	3.5 days	1.9 (1.3)	3.0 (2.4)	2.6 (2.3)
Perch	3.5 hours	0.7 (0.4)	1.1 (0.6)	1.2 (0.9)
	3.5 days	1.0 (0.4)	1.3 (0.9)	1.2 (1.0)
Roach	3.5 hours	0.4 (0.2)	0.6 (0.3)	1.1 (0.5)
	3.5 days	0.7 (0.2)	0.7 (0.7)	1.4 (1.0)

Table 1.13: Interval of concentrations of dissolved oxygen resulting in total fish population mortality as a function of duration of exposure and temperature (from Dowing and Merkens, 1957, ref. 2)

In addition to the problem of low concentration of oxygen there are risks associated with the presence of ammonium. This compound of nitrogen is typical of waste waters and is toxic in the form of ammonia. Transition to this form depends on the pH and temperature. The process of conversion from ammonium to ammonia becomes crucial when the value of pH exceeds 8.4. Such conditions are possible in a eutrophic environment in which photosynthesis may lead to high values of pH.

Consequently a flux resulting from a storm event has a reduced impact compared to the same flux resulting from waste water. The shorter the event, the higher the reduction of impact.

4.2 DELAYED OR LONG-TERM EFFECTS

Delayed or long-term effects are more difficult to comprehend because of the nature of contaminants consisting of heavy metals, hydrocarbons and other organic microcontaminants. The major characteristics of pollution with delayed effects are the persistence and accumulation of a large number of contaminants within the alimentary chain.

Figure 1.9, taken from Whitelaw, cited by **Ellis** (ref. 11), shows the delayed effect of Zn on species of *Gammarus* and *Asellus* due to accumulation of this metal in their tissues. Like others, this author too uses various species for characterising the effect of such cumulative contaminants.

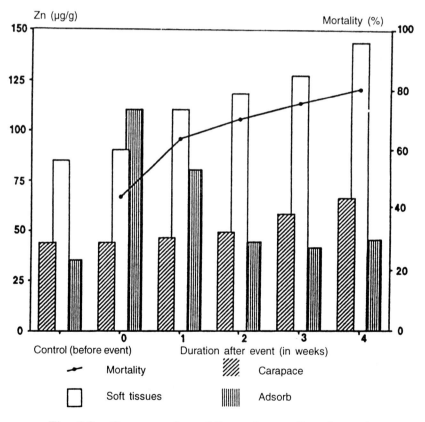

Fig. 1.9: *Concentration of Zn and mortality of certain invertebrates after storm events.*

Although it is difficult to compile the effects of these contaminants on the natural environment in which synergism and antagonism exist among various contaminants, necessary steps must be taken to reduce their concentration.

4.3 EXAMPLE OF IMPACT OF RUN-OFF WASTES IN A PARISIAN REGION

The reader is referred to case study 1 (in the second part of this book) edited by M. Mouchel and presented at the seminar 'Storm Weirs of Achères' in May 1995. It contains an analysis of the effects of discharge into the Seine River from the weir of Clichy, the largest among the Parisian conglomerate.

It is shown that the model of Streeter and Phelps underestimates the deficiency of oxygen for a small discharge and overestimates it for a large discharge and, further, that the presence of phytoplankton in the river causes much larger deficiences for the same discharge.

For a better understanding of these mechanisms which cause such differences, the author proposes two types of explanations of the phenomena:
— first, sedimentation in the Seine and the interaction between loose deposits and the water column,
— secondly, evaluation of the quantity of heterotrophic bacteria of the wastes and the biodegradability of organic matter.
These ideas will be taken up again in the programme 'PIREN' Seine.

CHAPTER 2

Various Types of Weirs and their Accessories

1. STORM WEIRS AND BYPASSES (SEPARATION OF EXCHANGE WORKS)

• In a combined network, **a storm weir** is defined as the ensemble of devices whose function is to discharge peak flow into the natural environment to prevent overload in the downstream network.

A second function of the weir is **to ensure a diversion of flow of contaminants** between the natural environment and the downstream sewer; the diverted fraction may differ from the hydraulic volume, depending on the type of weir, its location in the network and the auxiliary works.

• This weir may also divert surplus water during storms to another sewer and thus play **a role in making full use** of the capacity of the transport systems; such a device is called a **bypass** or **exchange work**.

• Another bypass work, called the **separator bypass**, situated on a combined network, similar in design to a suction device, performs a different function, namely diversion of water during dry periods to a waste water sewer. Thus, it transforms the downstream part of the combined sewer to a semi-rainstorm sewer. Obviously, this is only possible with a device equipped with one or two valves.

1.1 COMPONENTS OF STORM WEIR AND BYPASS

A storm weir always comprises:
- **a diversion work;**
- **a channel or sewer for diverting** 'overflow' to a natural drain (stream, river), including the discharge work itself on the drain. This channel (or eventually a channellisation) can be very long.

The storm weir is conected (Fig. 2.1)
- on the upstream: to a sewer bringing the combined water;
- on the downstream: to an outlet sewer which diverts waste water to a treatment plant.

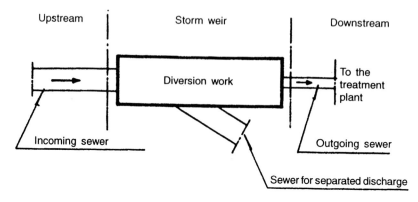

Fig. 2.1: Illustration of principle.

The diversion work may comprise all other components of a so-called hydraulic weir.

A system of storage (storm basin) may be connected to a diversion work for temporarily storing a part of the flood (especially the first storm flood).

Figure 2.2 schematically shows additional works found in the most complete weirs. Components 5 to 13 are described in this chapter while components of the storage/decantation basin are described in Chapter 4.

A bypass diverting water from a full sewer to another and functioning as a division work is the simplest and does not include a storage basin.

A sketch of such **a bypass** is shown in Fig. 2.3. The inflatable dam can be replaced by a main valve which opens during a storm while the secondary valve closes. Once the storm has abated, the flow of waste water is diverted to the network by opening the secondary valve and closing the main one (see Sec. 3.2).

1.2 DIFFERENT TYPES OF WEIRS

The only component that characterises a storm weir is the diversion work, of which there are three distinct types:
— works with overflow weir,
— works without overflow weir,
— works with vortex.

1.2.1 Works with overflow weir

In these works, an overflow weir is provided which diverts part of the flow when it exceeds a certain level. The objective is to limit the flow towards a treatment plant up to a specified value and divert the balance.

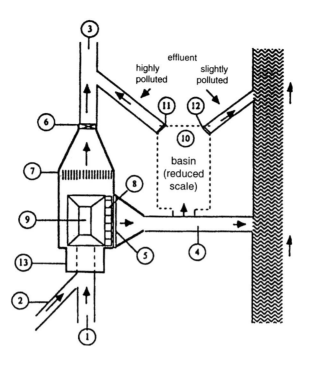

1. Combined effluent
3. To treatment plant
5. Overflow
7. Mechanised screen
9. Sand or rubble trap
11/12. Restoring valves

2. Lateral inflow
4. To natural environment
6. Remote-control adjustable valve
8. Harrow
10. Buffer storage basin
13. Settling chamber

Fig. 2.2: Detailed sketch of a storm weir.

To maintain the flow rate at a value which, if exceeded, would set the weir in operation and to limit the flow towards the treatment plant, a constriction is provided on the outgoing sewer (mask or section of reduced diameter, called the 'constricted section'), which allows calibration of the overflow weir at a higher setting.

According to the presence or absence of this constriction, overflow weirs are categorised into two main types:

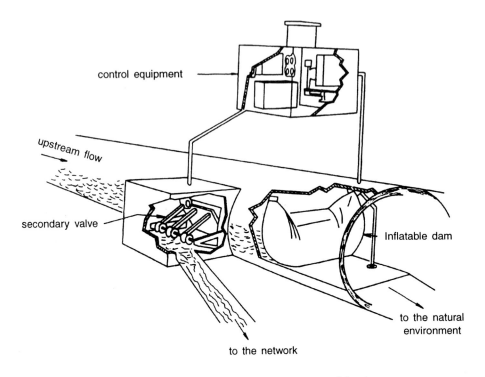

Fig. 2.3: *Separator bypass with inflatable dam.*

• **Weirs with high setting** (Fig. 2.4)

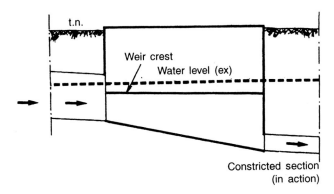

Fig. 2.4: *Weir with high setting.*

- **Weirs with low setting** (Fig. 2.5)

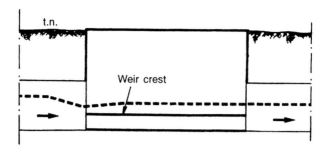

Fig. 2.5: Weir with low setting.

Depending on their orientation with respect to the sewers, their shape and that of the connections at the inlet and outlet, weirs can be classified as:

- **Weir with lateral overflow:** A weir of this type can be placed either on just one side (Fig. 2.6) or on both sides of the work. In the section corresponding to the overflow weir, the inlet pipe may be continuous or convergent.

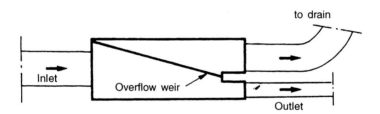

Fig. 2.6: Weir with lateral overflow.

- **Weir with frontal overflow:** The outlet sewer may be located along the axis of the inlet sewer, i.e., under the overflow weir, and may be directed laterally. This holds for the case of weirs with a high setting level (Fig. 2.7).

The overflow weir, generally horizontal, may be constructed at one level or be 'stepped'.

- **Advantages and drawbacks of works with overflow weir:** The best solution for limiting the flow rate in the downstream sewer consists of blocking a section of this sewer.

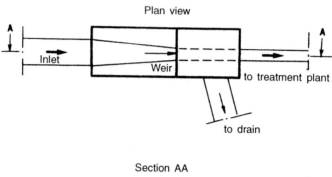

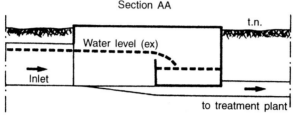

Fig. 2.7: Weir with frontal outflow.

Under these conditions, the flow during a dry period normally transits in the downstream sewer. When the flow rate reaches the limit allowable downstream or slightly less, the water level is the same as that of the overflow weir. If the flow rate increases beyond this value, the excess water flows above the level of the weir. As the water level at the weir is always relatively low, the variation of discharge over the weir and hence of flow rate in the downstream sewer is very small (because it is proportional to the square root of the discharge).

A weir with a high setting helps to maintain a low velocity of approach, to settle the flow and enables better knowledge of the hydraulic conditions prevailing.

The operation of this system is thus simple and clear; its computation (see Chapter 3) is also easy. This system is selected when the flow in the sewer at the inlet is in a laminar regime or when the projection change in flow corresponding to passage from the turbulent to the laminar regime, can be allowed upstream of the work.

Another advantage which will be elaborated later, concerns the division of the polluted flow (Chapter 4). It directs a major fraction of this flux to the plant, especially the recovery of deposits upstream, which often takes place before the weir overflows.

With respect to the weir with a low setting, due to the lateral opening of the sewer, conditions of hydraulic flow prevail from the upstream to the downstream and the overflow varies strongly with the slope of the floor of the sewer.

The water surface at the weir may in fact be of different configurations (depth of water at the head of the weir may be less than that at the extremity, or vice versa, and the projection may be upstream, downstream or in-between). There are thus a large number of possible cases of operation, more or less well known, which explains in part the large number of formulas proposed by various researchers, which are sometimes mutually contradictory (see Chapter 3).

The other drawback pertains to the pollution added to the environment, which, due to the recovery of deposited material, can be considerable.

1.2.2 Other discharge works

The oldest, now abandoned because of poor hydraulic efficiency and deplorable balance of pollution, are the **holes in the wall** (Fig. 2.8), with provision for improvement by advancing the opening (shown in broken lines) of discharge to the river.

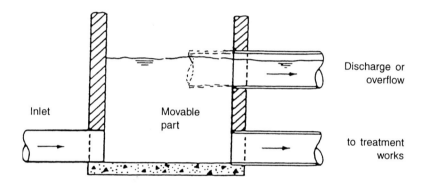

Fig. 2.8: *Wall-hole type Weir.*

Alternately, a hole **may be provided at the bottom** (Fig. 2.9).

Quite a large number of such works still exist in the present networks and should be replaced.

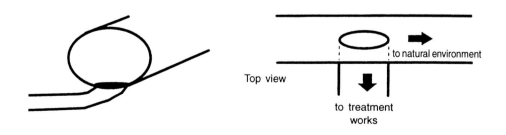

Fig. 2.9: *Bottom-hole type Weir.*

27

• **Siphons:** Their use has grown with a system of simple and efficient regulation. A controlled partial siphon has been installed at Seine Saint Denis (see District of Seine Saint Denis [69]) (Fig. 2.10).

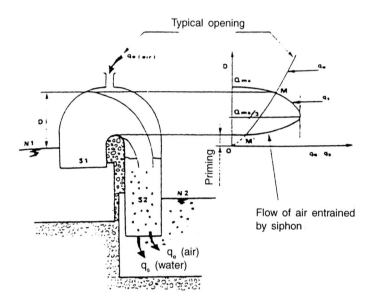

Fig. 2.10: *Sketch of operation of a partial siphon.*

An automatically controlled pneumatic valve makes it possible to partially compensate the flow q_s of entrained air by the fall of water at the rate q_e. By thus controlling the depression at the top of the siphon, the flow rate of water is adjusted. Automation is effected by pneumatic or electronic systems using standard equipment. The advantages of this system are **compactness**, very useful when the subsoil is cluttered, **and reliability**, since the moving parts and the control equipment are enclosed in a chamber.

• **Weirs equipped with adjustable valves**
The primary function of a valve is to replace the beams which were earlier used to adjust the weirs; it further helps in changing a weir with a low setting to one with a high setting. Such a transformation assumes a rigorous adjustment for preventing a high setting as well as a very low setting downstream which can cause frequent overflow.

This leads to a new selection of equipment with valves which can be adjusted according to the levels in the sewer and the river (Fig. 2.11).

28

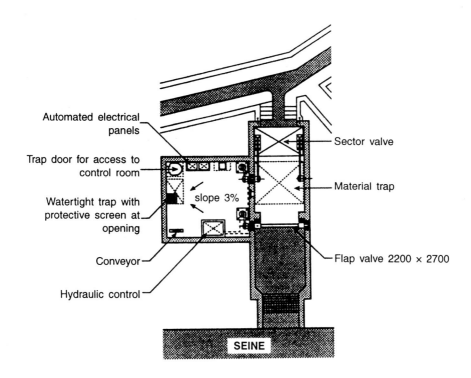

Automated electrical panels

Trap door for access to control room

Watertight trap with protective screen at opening

Conveyor

Hydraulic control

slope 3%

Sector valve

Material trap

Flap valve 2200 × 2700

SEINE

Fig. 2.11: *Equipment for an overflow weir.*

The equipment of these overflow weirs often includes a sector valve near the top. It is operated by a hydraulic control with an automated command transmitting the following information: level in the sewer as well as in the river. Safety in the case of flood is ensured by a flap valve. The commands given by the automatic system can be modified by remote control, since the position of the flap valve and the levels are communicated to the central station, which can then compute the quantity evacuated. A device of this type helps in introducing supplementary data in the command—quality of water in the sewer—enabling adjustment of the opening for less contaminated water.

Constant (case study no. 2) has elaborated the policy regarding equipment of the 20 largest weirs of Paris and described the functioning of the site of Alma and its management.

Table 2.1, taken from the **'Guide to Valves'** (ref. B), lists the factors on which depend the choice of type of valves most suitable for the weir, with due consideration of the site.

Vertical valves are often used for the bypass while sector valves are more suitable for weirs. Figure 2.12 describes sector and flap valves, both used for

29

Type of valve / Criterion	Sector valve	Vertical valve	Horizontal valve	Flap valve	
				Jointed	Non-jointed
Isolation	xxx	xxx	xxx	x	x
Control	xxx	xx	–	xxx	x
Cluttering					
—vertical	xxx	x	xxx	xxx	xx
—horizontal					
along direction of flow	x	xxx	xxx	x	x
in direction perpendicular to flow	xxx	xx	0	0	xxx
Silting of weir	xx	xx	0	x	x
Inspection accessibility	xxx	x	0	x	x
Maintenance	xxx	xxx	x	x	x
Investment	High	Medium	Medium	Medium	Medium
Operational cost	Medium	Low	High	Low	Low

Legend: xxx   Very good
xx   Good
x   Medium
0   Bad

Table 2.1: *Criteria for selection of valve type*

regulating the level in sewers and on the weir. The **'Guide to Valves'** (ref. B) gives a detailed comparison of advantages and drawbacks and provisions to be made in civil works for each type of valve.

• **Inflatable dams:** The adjustment of a weir by an inflatable rubber dam which can conform to the shape of a sewer is depicted in Fig. 2.13. The dam is more stable and has a very large mechanical inertia. It has great importance in large civil works but is also useful for constructing dams with no need for reconstructing the sewer to install a valve in it.

SECTOR VALVE (or valve of overflow weir or of storm weir)

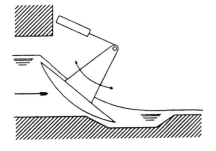

This valve moves about its joint and is activated by a jack. It is primarily used for regulating the flow rate or a water body or over a weir which can be readily adjusted to predetermined positions.

FLAP VALVE WITH LOW JOINT

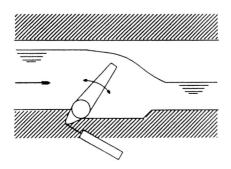

This valve is activated by a jack and moves about its joint. It is primarily used for regulating the flow rate in a water body or over a weir, which can be readily adjusted to predetermined positions.

FLAP VALVE WITH HIGH JOINT

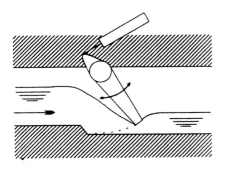

This valve is activated by a jack and moves about its joint. It is used for isolating or controlling a maximum flow rate. It is recommended for situations wherein the depth is too small for using a vertical valve.

Fig. 2.12: Three types of valves suitable for weirs.

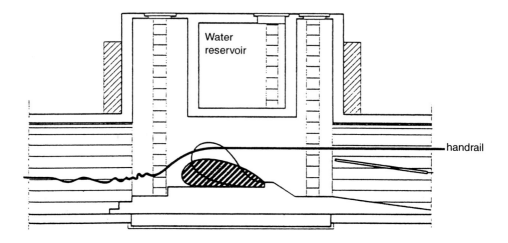

Fig. 2.13: Inflatable dam.

1.2.3 Works with vortex

The dimensions of a vortex chamber with peripheral weir derived from the 'Swirl Concentrator' of 1970 (known in France as the Static Vortex Separator) are given in Fig. 2.14.

The purpose of these works is to reduce the kinetic energy of flow to help deposition of particles that remain suspended due to the length of the trajectory which is roughly helical. Besides, the rotational flow produces secondary centripetal flow near the bottom and collects the particles which decant towards the bottom.

These works will be discussed in Chapter 4 and their efficiency of operation evaluated. Their complexity, and therefore cost, is indeed compensated by the advantage of pollution reduction which they ensure.

2. ADDITIONAL OR 'COMPLEMENTARY' WORKS OF WEIRS

2.1 SCREENS (AND MESHWORK)

These are meant to retain gross solid particles ($d > 6$ mm) in the weir and thus preclude their transmission to the natural environment. In fact, the presence of these particles, mostly floating, is very unaesthetic and for the public, synonymous with pollution. To be effective **these screens should be equipped with means of automatic cleaning** so that the particles are stored and later readily removed

SECTION

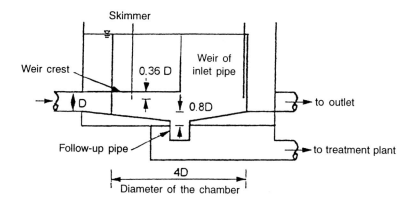

PLAN

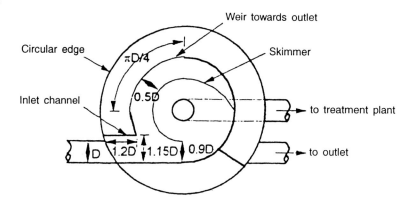

Fig. 2.14: Vortex chamber with peripheral weir.

in a preplanned manner. In many weirs, **constraints of cluttering and access make installation of such screens very difficult.** For these reasons, screens are sometimes placed upstream of the inlet sewers or even on the siphons and settling basins, thereby promoting the systematic trapping of particles upstream.

For example, at Seine Saint Denis there are only 11 screens and none is attached to the weir; they are placed at the head of the storage-settling basin or on uncovered rainstorm sewers. They remove 1000 tons/year of wastes, i.e., 0.8 kg/inhabitant.

The bars of these screens are 10 to 12 mm thick and generally spaced 10 to 15 mm apart. The mechanical rake inserted in this space needs a play of 3 to

33

4 mm; the teeth are 6-7 mm in size, otherwise the rake would lose the rigidity of the screen bars necessary for proper operation. So it is impossible to provide smaller than 10-12 mm spacing between bars. Such screens have an efficiency of the order of 50% (see **Riding** [20]).

Researchers in England are currently experimenting with fixed or adjustable screens with smaller spacing (about 6 mm), which are more suitable according to the new English legislation, which gives priority to the removal of solid particles larger than 6 mm. Their cost is higher due to a system of cleaning by brushes or by a rotary device.

Meshes with spacing of 4 mm have been used in Germany and Switzerland for the past several years. According to **Fahrner** [75], they prevent the inflow of solid particles larger than 4 mm in rivers, underground basins etc. These screens can be fed from bottom to top or top to bottom when they are horizontal, or transversely when vertical; these arrangements are the most suitable. The screens cause a loss of energy equivalent to a few centimetres but need an input of energy equivalent to about 20 centimetres; hence it is necessary to lower the weir. However, a deflector can re-establish the overflow on the initial side (Fig. 2.15).

The material retained by the mesh is diverted to the treatment plant; the flow of water cleans the screen.

The district of Hauts de Seine (see **Soulier,** case study no. 6) is trying out a floating dam, 'raft', at Pont de Sèvres, which constitutes a siphon wall guided by two sliding channels conforming to the section of the sewer; a few exploratory trials have already been carried out.

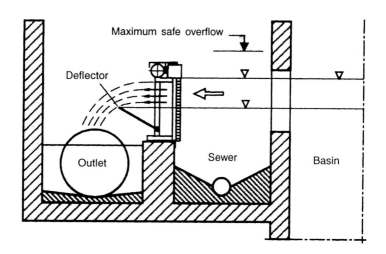

Fig. 2.15: Installation of mesh on a weir,
either on a sewer or on a basin.

34

2.2 Settling and degritting chamber

The purpose of this chamber, located upstream of the weir, is to reduce the flow velocity to facilitate decantation of the heavier suspended particles and to raise flotsam to the surface.

Figure 2.16 shows a chamber of this type connected to a frontal overflow weir which retains the flotsam and despatches it to the treatment plant after a storm. To prevent silting, **Balmforth** [59] recommends:

$$D_{min} = 0.815 \, Q_p^{0.4}$$

where Q_p in m³/s represents the maximum flow rate during the 2 to 5-year return period and D_{min} is the minimum diameter in metres.

It is also advisable to have a minimum slope of 4/1000 over a length of at least 25 D upstream of the weir; this ensures recovery of the deposited and buoyant matter when the water level drops.

A settling chamber connected to a lateral weir with two setting levels is depicted in Fig. 2.17.

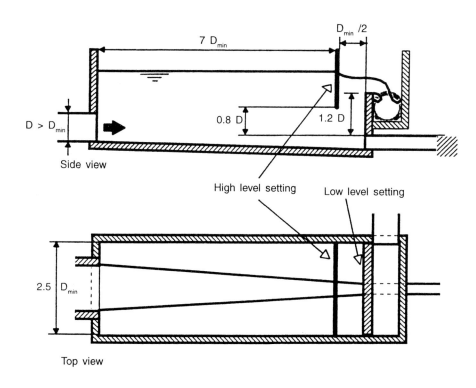

Fig. 2.16: *Dimensioning of a frontal weir with a settling chamber.*

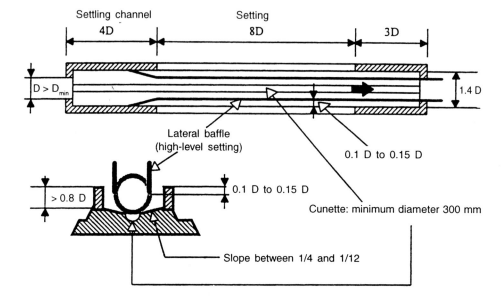

Fig. 2.17: *Rules for dimensioning lateral weirs with a two-level setting.*

This particular shape of a lateral weir whose dimensions were determined through studies on reduced models carried out in the early 1980s (**Saul and Delo** [13]) has the following advantages:

— long and narrow settling chamber enabling larger particles to be decanted and flotsam to rise to the surface;

— constriction at the outlet regulates flow and precludes formation of a swirling flow in the chamber;

— lateral baffle acts as a siphon partition and retains flotsam.

It should be noted that in Fig. 2.17 a 'lateral baffle' has been used on the weir with a high-level setting instead of screens; at the time of drop in water level the flotsam is thereby diverted to the decanting screens downstream or to the treatment plant.

2.3 WATER RISE IN CASE OF FLOOD OR HIGH TIDE
(SEE AGHTM [REF. C])

For sewers located in low zones and along rivers, the overflow works should be provided with pumps to pump out the water in sewers, thereby ensuring that during floods part of the overflow will be transferred directly to the river.

The number of such pumps can be fewer than the number of weirs on low sewers since pumping will eventually reverse the direction of flow on the section of the sewer situated downstream. Special precautions should be taken for degritting contaminants and, within a city, foul air should be treated to remove odours. A recently installed pumping station for waste water and rainstorm run-off in case of flood is shown in Fig. 2.18.

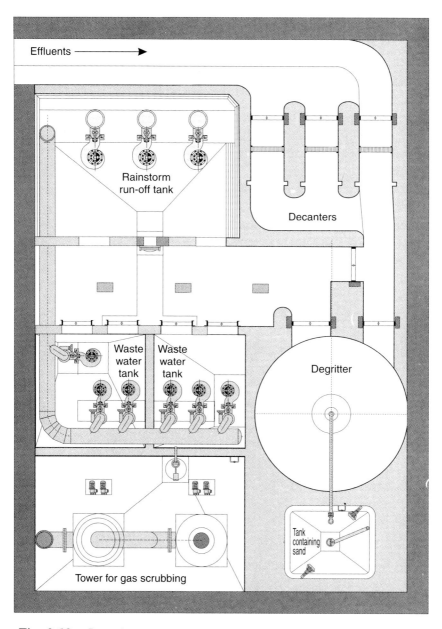

Fig. 2.18: *Pumping station for waste water and safety against flood.*

The effluent travels from upstream to downstream:

— **decanters:** three identical machines are provided; two suffice for treating the discharge of the project, the third is kept in reserve in case a machine breaks down;

— **degritter:** this equipment comprises the degritter (with tangential inlet), an intermediate tank for storing water containing sand and a screw held in the sandy suspension by pumps placed in the recovery bank. The degritter can be completely isolated by valves in the partition wall;

— **two identical sets of pumps for waste water,** each closed by a pair of valves. If this system cannot handle the quantity to be pumped, the number of pumps may be doubled. Further, an independent plant for generating electricity should also be provided.

The set of pumps for rainstorm run-off are placed near the set of pumps for waste water and activated as soon as the excess flow causes a rise in level of the water body.

The air of the pumping station is deodorised by counter-current scrubbing before discharge into the atmosphere.

The pumping station shown in Fig. 2.18 is equipped with water-level indicators: mercury bulbs are used only for spotting very high or very low levels.

For example, in Hauts de Seine there are 13 pumping stations for 66 weirs situated along the Seine River (including that of SIAAP at Clichy); the quantity pumped is 35 m^3/s (likely to be raised by another 10 m^3/s) added to that at Clichy.

In estuarine towns, such as Bordeaux, Rouen or Le Havre, these stations pump waste water round the clock during high tide. For example, there are 75 stations along the perimeter of the CUB of Bordeaux for pumping 65 m^3/s, which is more than the minimum flow of Garonne.

2.4 BASINS ASSOCIATED WITH WEIRS

Figure 2.19 enables understanding the operation of these basins which can be divided into two types: off-line and on-line.

• **Off-line basin** (sketch 1)
Out of the direction of flow, this basin is fed by weir S_1 which diverts that flow to it which cannot be taken by the downstream sewer. When full, water is evacuated to the natural environment by storm weir S_2.

• **The on-line basin** is fed directly by the entire water, a weir diverting to the natural environment the flow that cannot be taken by the downstream sewer when the basin is full (sketch 2). The flow downstream is often regulated by a valve. Sometimes the weir is located upstream (sketch 3).

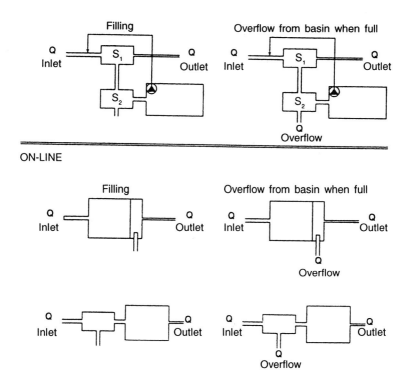

Fig. 2.19: *Sketches illustrating the principle of off-line and on-line basins.*

The outflow is integrally treated in the treatment plant; it receives the transitory flow and later the flow from evacuation of the basin which often has to be pumped, as in the preceding case.

These basins have a twofold effect:
— separation which reduces the quantity and frequency of overflow;
— decantation related to basin storage time and geometrical configuration of the basin which reduces, in the case of transit, the quantity of pollution that overflows when the basin is full.

The two types of basins can be combined to take advantage of trapping the eventual first flood when it occurs (see Chapter 4) and partial decantation of contaminants.

Chapter 4 contains the modalities of computation of these basins while Table 2.2 summarises the specifications, advantages and drawbacks of the three types— on-line, off-line and combined.

The choice is based on financial considerations—comparison of efficiency vis-à-vis additional cost. It would seem that when the more polluted first flux

On-line basin	Off-line basin	Combined basins
Specification —Evacuation by gravity	—By gravity or pumping	—Suitable for pumping
Advantages —Hydraulic simplicity —Lower cost —Smaller ascendary	—Ease of cleaning	—Ease of cleaning —Maximal efficiency of decantation
Drawbacks —Cleaning difficult due to perpetual presence of water in basin except in case of channellisation of storage	—More complex than on-line and thus cost higher —Ascendary extended	—Most complex and thus most expensive —Ascendary extended

Table 2.2

appears over a weir in a network, two basins is the better solution. According to some researchers, this is true of weirs for which the time of retention of storm water is shorter than 10 minutes (this does not happen when t > 20 minutes), and finally when the flow velocity during dry periods lies between 0.5 and 0.8 m/s (see Chapter 1).

3. SOME CONCLUDING REMARKS ON WEIRS AND BYPASSES

3.1 WEIRS

Originally conceived for serving only a hydraulic function, under appropriate conditions weirs can discharge two other important functions:
— reduce the concentration of polluted flux discharged in the natural environment compared to that of the downstream flow towards the treatment plant;
— allow partial storage of the storm flux in the upstream sewer and hence a reduction in frequency of overflow.

The function of partial separation of the pollution is discussed in Chapter 4; the discussion includes not only the weir but its accessories, in particular the associated storage/decantation basin.

The effect of storage in the network induced by a weir is not very significant but nonetheless cannot be ignored. For networks in flat zones, e.g. The Netherlands,

40

storage can be as high as 70 m³ per impermeabilised hectare; it reduces with the slope but is never less than 20 m³. However, let it be noted that management is complex, given the risk of overflow. This reinforces interest in weirs with high settings and such devices as control valves etc.

3.2 BYPASSES

In **a bypass with a mesh** the function of concentration of contaminating flow is not important but that of increasing storage in the network (and thereby reducing volume stored in the basin) is. Equipment such as valves of large weirs and computerised management of the systems merit attention.

The same holds for **a suction bypass** which diverts the flow during dry periods to a network of waste water.

A sketch was given earlier (Fig. 2.3) of an automatic device with remote control used in the **Valerie** system [71] to assist in the organisation of the large combined sewers of the Parisian region for diverting part of the water of the dry period towards Valenton. The objective of the district of Val de Marne (see **Pister** [58]) is twofold:

— reduce the contribution to Achères in order to prevent further expansion of the plant, which had already grown to a gigantic size;

— reduce the discharge during the rainy period by fully utilising the capacity of sewers capable of storing 300,000 m³ of waste water diverted to Valenton. The overflow from the annual rainfall is reduced by about 100,000 m³.

The other significant fall-out is the improvement of knowledge of the quality of effluents transiting in tanks on which suction is effected. Improvement has been a result of measurements taken over a period of five years at ten locations (see **DEA Val de Marne** [65]). A device for continuous measurement of pollution has now been installed (see **Rabier**, case study no. 4).

CHAPTER 3

Computation and Measurement of the Charateristics of Weirs

Computation and measurement are complementary both for the installation and management of weirs. However, measurement is the only means for evaluating the efficiency of these works with respect to separation of contaminating flow.

1. COMPUTATION OF WEIRS

1.1 CHOICE OF TYPE OF WEIR

The characteristics of a given work for a predetermined discharge mandate selection of the right type of weir.

The basic data necessary for computation of the rate of flow into an upstream sewer comprise:

If Q_{DS} is the flow rate during the dry season and Q_{max}, the maximum flow rate allowed in the upstream sewer, while Q_{lim} is the limiting flow rate allowed in the downstream sewer, the overflow rate Q_{over} is given by:

$$Q_{over} = Q_{max} - Q_{lim}$$

This information must be supplemented with the following data concerning the sewers, taking into account the mean slope:

— diameter of the upstream and downstream sewers under gravity flow for the largest flow rate planned in each;

— flow conditions (depth, speed, type of flow—streamline or turbulent) for various flow rates;

— average available power for discharging water into the drain as well as the possible zone of installation of the storm weir together with the topography of the terrain and inventory of obstacles (cables, conduits etc.).

It is generally possible with this information to either directly determine the type of storm weir to be selected or to make economic comparisons of pre-project summaries of each potential solution.

As a matter of fact, the choice is limited in most cases to overflow weirs (with high or low level setting) or to simple openings in the floor. Given their complexity or constraints in operation, other systems are employed only in very special situations.

If the flow is streamline and not large above the drain, an overflow weir is the best choice.

1.2 COMPUTATION OF OVERFLOW

For dimensioning the work it is necessary to calculate the overflow per running metre of weir in terms of depth of the overflowing layer.

• **The general formula of the law of overflow in a rainstorm regime for a lateral weir** is as follows (Fig. 3.1):

$$Q_{over} = m \, L \, h \, \sqrt{2gh} \qquad (3.1)$$

where m is the coefficient of discharge; L, the weir length; and h, the depth, upstream (h_0), downstream (h_1) or in the middle, as the researcher chooses.

The coefficient of discharge depends primarily on depth Z, of invert of channel below the weir crest, velocity of approach, obliquity of the weir and ratio of overflow rate Q_{over} to inflow rate Q_{max}.

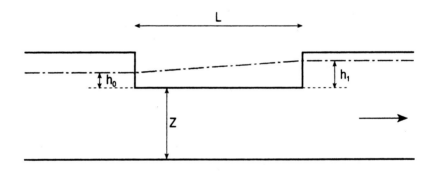

Fig. 3.1: *Lateral weir in streamline flow regime.*

Applying **Dominguez'** formula

$$Q = m \, L \, h \, \sqrt{2gh} \qquad (3.2)$$

it is possible to calculate Q from **the average depth h** and the shape of the weir crest using Table 3.1 if the difference between h_0 and h_1 is very small. If h_0

and h_1 differ significantly, the coefficient of discharge m should be multiplied by a coefficient k given in Table 3.2.

Crest shape	Average depth h in metres					
	0.10	*0.15*	*0.20*	*0.30*	*0.50*	*0.70*
Thin crest free fall	0.370	0.360	0.355	0.350	0.350	0.350
Broad crest	0.315	0.320	0.320	0.325	0.325	0.330
Broad rough crest	0.270	0.270	0.273	0.275	0.276	0.280

Table 3.1: Coefficient m

	Streamline regime $h_0 > h_1$	k
h_1 / h_0	= 0.4	0.598
	= 0.5	0.659
	= 0.6	0.722
	= 0.7	0.784
	= 0.8	0.856
	= 0.9	0.924
	= 1	1

Table 3.2: Value of k as a function of h_1 / h_0

Various possible configurations of flow along the weir are shown in Fig. 3.2 and Table 3.3 taken from the study of the Ministry of Research [ref. D] for streamline or turbulent flows.

Type of flow	Slope along weir	Downstream slope	Flow conditions
Case (a)	Slight	Slight	$h_{nl} > h_{cl}$ and $h_s < h_{cl}$
Case (b)	Slight	Slight	$h_{nl} > h_{cl}$ and $h_s < h_{cl}$
Case (c)	Slight	Slight	$h_{nl} < h_{cl}$ and $h_s < h_{cl}$
Case (d)	Steep	Steep	$h_{nl} < h_{cl}$
Case (e)	Steep	Slight	$h_{nl} < h_{cl}$
Case (f)	Steep	Slight	$h_{nl} < h_{cl}$

Table 3.3

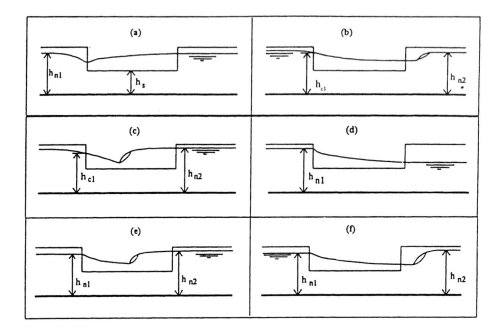

Fig. 3.2: *Various flow configurations on a lateral weir*
with : h_{n1} : the normal depth in the upstream conduit
h_{c1} : the critical depth in the upstream conduit
h_{n2} : the normal depth in the downstream conduit.

Chocat [14] has given the mode of computation suitable for long weirs for the two types of flow, especially with the program 'Overflow' of **Mitri** and **James** [15].

In the simplest case of a streamline regime with $h_0 \neq h_1$, the overflow is determined by an iterative computation scheme starting with the downstream flow rate Q_{down} corresponding to depth Z. The overflow rate in the first iteration (using formula 3.2) is given by:

$$Q_{over1} = Q_{max} - Q_{down1}$$

In the second iteration, the depth is taken as $h + Z$ and the new value $Q_{over 2}$ is obtained. This iterative procedure is continued until one obtains

$$Q_{over\ n} \cong Q_{over\ (n+1)} \quad \text{or} \quad Q_{down\ n} \cong Q_{down\ (n+1)}$$

Convergence is much faster for a weir with a high setting level because the variation in h results in a very small change in Q_{down}.

46

- **Frontal weir with high level setting**

Figure 3.3 shows the major dimensions of a weir with a high level setting (height above invert = Z, depth above weir = h, constriction = ∅D and length = l).

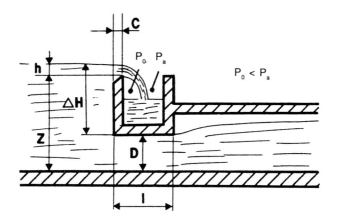

Fig. 3.3: Frontal weir with high level setting.

The downstream flow rate before overflow (h = 0) corresponds to a depth $H = h + Z - D_1$ and is given by eqn (3.3) where A is the sectional area of constriction, λ the loss coefficient and R_h its hydraulic radius:

$$Q_{down} = V_{down} \times A \quad \text{with} \quad V_{down} = 2 \left(\sqrt{2g\Delta H R_h / \lambda l} \right) \quad (3.3)$$

Corresponding to an upstream flow rate Q_{up}, the overflow is therefore given by

$$Q_{over} = Q_{up} - Q_{down}$$

The flow rate of the weir is given by a formula of type (3.1) which helps in computing a first value h_1. This value modifies H and gives, through application of eqn (3.3), a new value $Q_{down\ 1}$, which slightly reduces Q_{over} to the value $Q_{over\ 1}$. This process is repeated until a value h_n is obtained such that $h_{n+1} \cong h_n$. Convergence is fast if h is small compared to [Z – D].

The simplicity of these formulas is explained by the fact that the weir is perpendicular to the approach velocity. Table 3.4 lists three formulas for thin weirs[1]; those of **Bazin** and **Rehbock,** with their limits of application, are more widely used. Document 1 of Appendix I gives the values of m for each of these formulas when the weir is of the same length as the inflow channel; it also includes tables giving the flow rate in litres/s/m weir.

---

[1] A weir is said to be thin if $h_0 > 2c$. **Chocat** has given formulas for thick weirs and for the case when the pressures P_0 and P_a are unequal.

Author	Formula	Conditions of application
Bazin	$Q = m\, L\, h_0\, \sqrt{2g \cdot h_0}$ $m = \left[0.405 + \dfrac{0.003}{h_0} \right] \times$ $\times \left[1 + 0.055 \left(\dfrac{h_0}{h_0 + h_s} \right)2 \right]$	– Upstream channel rectilinear, of large length, without device for regulating velocities; – 0.08 m $< h_0 < 0.70$ m; – $L > 4\, h_0$; – 0.2 m $< h_s < 2$ m
Rehbock	$Q - m \cdot L \cdot H_e\, \sqrt{2g \cdot H_e}$ $m = 0.4023 + 0.0542\, \dfrac{H_e}{h_s}$ $H_e = h_0 + 0.0011$	– Upstream channel of small length but contains device for regulating velocities (for example, a screen with bars); – $h_0 > 0.05$ m
Barbe	$Q = m\, L\, H_0\, \sqrt{2gH_0}$ $m = 0.418 + 0.0120\, \dfrac{H_0}{H_s}$	

Table 3.4: *Formulas for frontal weir [(h_0) on weir and h_s depth]*

These formulas are valid if the overflow is free, i.e., the pressure P_0 under the sheet is equal to the pressure P_a (Fig. 3.3). If the overflow is submerged (Fig. 3.4), Bazin's formula gives the following specific expression for m:

$$1.05\, [1 + 2\, h_{down} / h_s]\, [(h_0 - h_{down}) / h_0]^{1/3}$$

which replaces that for Fig. 3.4.

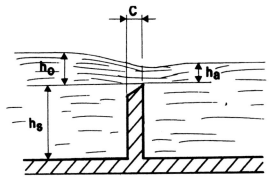

Fig. 3.4: *Submerged nappe.*

48

When the weir is shorter in length than the channel (lateral contraction), the following correction proposed by Hegly for the expression for m given by Bazin, incorporates coefficients α and β:

$$m = [0.405 + 0.003/h_0 \ \alpha] \ [1 + 0.55\beta \ h_0/(h_s/h_a)]^2$$
$$\text{with} \quad \alpha = 0.0027 \ (L - L_0) / L \quad \text{and} \quad \beta = (L_0/L)^2;$$
$$L \ \text{and} \ L_0 \ \text{are defined in Fig. 3.5}$$

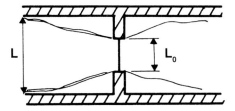

Fig. 3.5: *Weir with lateral rectangular contraction.*

• **Weirs with multiple (stepped) or inclined crest** (Fig. 3.6)

In the computation of such weirs, the same formulas are applicable for several weirs (one step at a time); alternatively, the inclined surface may be treated as consisting of several steps.

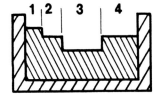 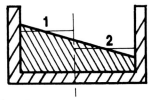

Fig. 3.6: *Stepped or inclined weirs.*

2. CALIBRATION (OR ADJUSTMENT) OF WEIRS (WITH LIQUID FLOW)

As mentioned in the preceding section, the formulas giving the flow rate of weirs in terms of depth of overflow are applicable to works of simple shape, rarely encountered, especially among old weirs.

The coefficient m occurring in formulas, even for simple cases, varies strongly with the shape (thickness of weir, obliquity and length, open surface or otherwise etc.); the differences are not important, however, in establishing a works project because the necessary enlargement and reserve for fixed weirs—a safety trench for moderating the curve of maximum flow to preclude overflow—can still be planned.

The difficulty in applying a formula is well illustrated by the semifrontal weir shown in Fig. 3.7

On the other hand, to acquire knowledge of the effects of works on the natural environment and for network management, it should be possible to estimate fairly accurately the rate and volume of overflow by continuously measuring a simple parameter, such as depth of water. This quantity Q_{over} (h) for large works is likewise indispensable for modelling the network.

Although an approximate knowledge of Q (h) suffices for works which have a marginal effect on the natural environment, an accurate knowledge of this quantity is imperative for major weirs (or for a sector valve, the discharge $Q(\alpha)$ as a function of angle of valve opening α).

For all weirs, the value h = h_d corresponding to the start of the overflow $Q(h_d) = 0$ must be known.

2.1 MEASUREMENTS

The methods of measurement differ from one weir to another depending on their specificity and the ease with which the facilities for measurement can be set up. Depending on the possibilities offered by the discharge sewer, the following five methods can be proposed.

• Long sewer enabling establishment of a uniform flow regime

Utilisation of Strickler's law

$$Q = K S R^{2/3} I^{1/2}$$

(where R is the hydraulic radius, S the wetted cross-sectional area, I the gradient) necessitates measurement of the depth in the sewer to calculate K and concomitantly measurement of the reservoir depth.

• Discharge sewer with uniform cross section allows compution of the flow rate by measuring the head loss using an appropriate law; two measurements of depth in the sewer and one over the weir are required.

• Sewer with discharge of short duration

Three types of measurements are possible:
- provisional installation of a flowmeter,
- computation of flow rate in a section by measuring its depth and the velocity either by a microwinch or an electronic or ultrasonic probe,
- installation of an overflow gauge on the downstream of a weir.

In all three cases, the measurements should be correlated with the measurement of depth over the weir.

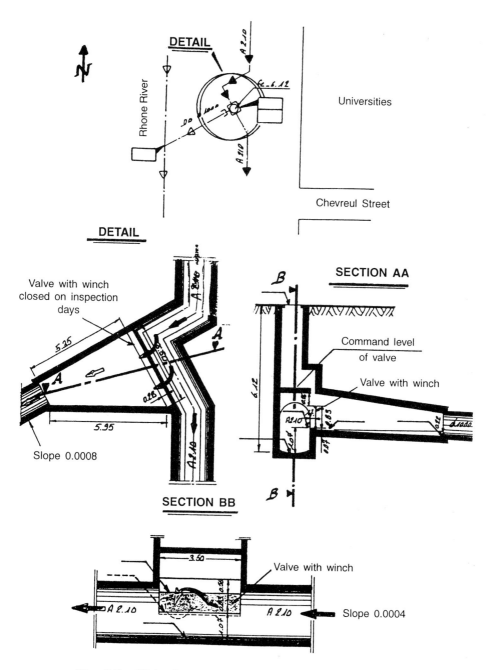

Fig. 3.7: *Weirs located at Lyon 7, Quai Claude Bernard (after Hodeaux [10]).*

• **Two measurements by sector gauges,** one upstream and the other downstream of the weir; the overflow computed from these measurements should be correlated with the measurement of depth over the weir. Taking the cumulative error into account, this method is suitable only for large overflows whose direct measurement in a short discharge sewer is difficult.

• **Selection of an appropriate standard** relating the depth and flow rate (for example by weirs equipped with valves or by conducting a model study, the latter justified only for large or complex works for which in-situ measurements are apparently not possible).

Choosing the best solution and the method of measurement is a tricky matter and consultation with a specialist is advisable.

It may be noted that the above-mentioned Q(h) law is not valid when the level H downstream of the weir exceeds the value H_1 and modifies the conditions of overflow (from an open surface to a submerged one, for example). Therefore it is advisable to determine the conditions of validity of the law for the case of flood and eventually determination of another law for $H_1 < H$.

Likewise, installation of a limnimeter giving the depth h over the weir must be done under the direction of a measurement specialist; the meter should be placed slightly upstream of the weir at a distance less than 4 l, l being the length of the weir.

In case study no. 3, **Béthouart** has described the approach used in the programme of measurements in 1993 and, in particular, specified the constraints whereby a decision was taken to select an overflow law or Strickler's law, or use the principle of loss of energy, or go for direct measurement—the choice depending, of course, on the type of weir.

Measurements in a discharge sewer (called a weir in Paris) should be carried out at a certain distance from the outlet to avoid turbidity and, if possible, at a point where river flooding can have no effect. Flood effects have invalidated measurements in a number of cases.

Experience has shown that measurements of depth should be repeated to ensure accuracy of values and the process can be readily automated.

2.2 Available material

• **Table 3.5 lists the characteristics of major level gauges.** Until recently, a bubble gauge was the most widely used but it requires maintenance and accurate calibration; besides, it is quite expensive. So, it has been supplanted with gauges based on measurement of stresses or employing the Wheatstone bridge.

More recently, ultrasonic gauges have appeared which enable very accurate measurement of liquid level without touching the liquid.

All these gadgets permit transmission of the data obtained over some distance.

	Electromechanical		Hydrostatic with membranes				Ultrasonic impulse type	Capacity type
	Floating type	Palpating type	Bubble type	Piezo-electric	Piezo-resistance	Capacity		
Design	Physical displacement of float or contact	B	Measurement of pressure	Stress on crystal	Stress guage	Surface of two condensers	Measurement of time of concentric propagation	Capacity with respect to armature test
Adaptation to sanitation	B	B	AB	Rarely used	B	B	TB	AB
Condition of installation	Axial pit	B	AB immersed	B immersed	B immersed	B immersed	B	B
Energy source	Sector	Batteries	Compressed air 3/6 months	Sector	Batteries 15 days	Batteries	Sector	Sector
Accuracy	M	TB	AB to average	B	B	B	B	AB
Reliability	TB	TB	AB	B	B	B	B	AB
Strength	TB	TB	B	AB	AB	AB	B	B
Maintenance	Easy	Simple	Adjustment easy, access probe at water level				Easy	Easy but adjustment difficult

Table 3.5: *Characteristics of major level gauges*

• Measurement of flow rate in network

The equipment used for this purpose measures both, depth at a section by means of **resistance or ultrasonic probe** (see Table 3.5) and velocity by electromagnetic or ultrasonic probe; the velocity and depth are computed by using appropriate software. One drawback of such equipment is its high cost, given which measurements are carried out only at a few crucial points; furthermore, its installation requires sensitive instrumentation[1].

When the flow in the sewer is uniform, the flow rate can be computed from depth at two points separated by a suitable distance. This requires measurement of flow velocity and evaluation of the friction factor. This is the method used by SIAAP [77] in the SCORE system which manages a group of drains ending at the Achères plant; it uses 185 piezoelectric level gauges and a computer software.

On the other hand, Fig. 3.8 shows a plan of continuous measurement in a sewer, consisting of depth and velocity at two levels of the same section of the sewer.

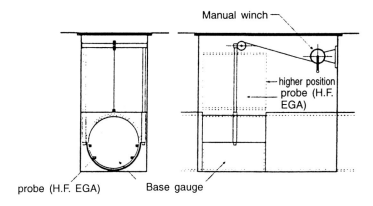

Fig. 3.8: Continuous measurement of flow rate in a sewer.

3. MEASUREMENT OF CONTAMINANTS

3.1 Facilities for measurement programme

Taken from a report of the LROP of 1989 [16], Fig. 3.9 shows a sketch of a measurement device installed on three weirs of the Bièvre sewer which comprises:

---

[1]See photo no. 15, document 1, Annex. I.

— a bubble limnograph (0-1 metre) and for safety two others, of variable amplitude (accuracy 0.5 to 1%);
— a completely programmed automatic sampler with a refrigerated tank;
— a triggering device to activate the sampler set at a predetermined position;
— a piloting device that guides the system and marks various operations (water level, sampling, labelling and counting samples [28 flasks possible]).

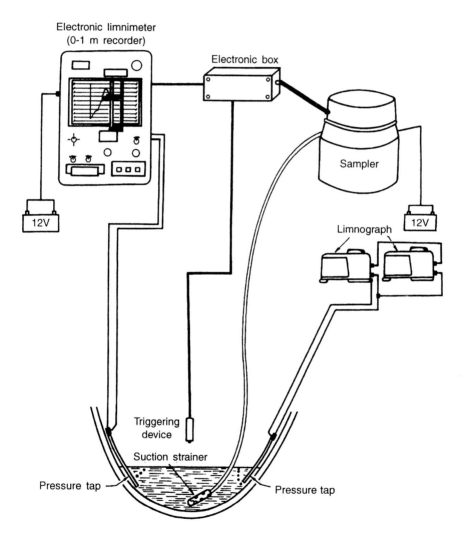

Fig. 3.9: *Station for measurement of contaminating flow.*

The device is deployed at the point where measurements are to be taken; the limnograph gives the flow rate, samples—4 per flask—at intervals of 5 minutes or spread over a 9-hour period.

The precautions to be taken during installation and choice of material are listed below:

— Careful determination in placement of the trigger so that the samples drawn are not from clean water from the bottom of the sewer.
— Determination of the level at which samples are to be sucked to preclude inclusion of silt from the bottom.
— Pumps should be of the peristaltic type, allowing sampling to be done at interval from 1 to 100 minutes.
— Flushing for a few seconds before sample drawing.
— Suction of an adjustable volume.
— Flushing after samples are drawn.
— Feeding rate of devices should be slow.

Laboratory analysis of the samples was done [results: BOD_5 - COD - MEST - total phosphorus - SO_4^- chlorides; and for average samples, heavy metals Cu - Pb - Cd - Ni and Kjeldahl nitrogen N + K].

These measurements enabled sketching hydrographs of overflow—variation in concentration and quantity of contaminants of different types. The M (V) curves for MEST for two weirs, drawn according to the procedure explained in Chapter 1, Sec. 2.2, are shown in Fig. 3.10.

These days the materials needed for sampling are available, as are limnographs (see Sec. 2.2). Nonetheless, care should be exercised to ensure that the discs on which the data are recorded and those used in the laboratory for analysis of information are compatible so that the various curves displaying results can be automatically traced. Some laboratories ought also to carry out exercises for error control and validation of measurements.

According to **Green** [20], analogous methods are employed in England and the material used for sampling is the same. However, the British recommend for a storm longer than 4 hours reduction of the interval of sampling to 10, 15 or 20 minutes, after 4 hours.

3.2 CONTINUOUS MEASUREMENT OF FLOW

3.2.1 Principles

For rainstorm water these measurements are currently based on the correlation between turbidity and type of pollution, mostly SS. Turbidity is measured by the amount of light absorbed, which depends on the grain size of the elements constituting SS. Other more complex and expensive devices can also measure COD (see Sec. 3.2.2). It has been shown (see **Grange** [17]) that the correlation

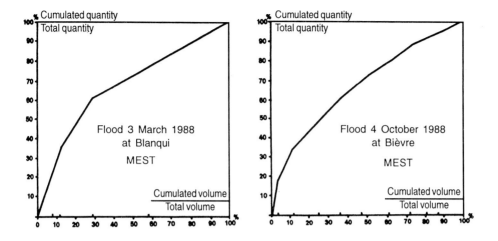

Fig. 3.10: *M (V) curves for MEST at Blanqui and Bièvre weirs.*

is better (r^2 increasing from 0.82 to 0.94) if a corrective term $1 + \alpha Q^B$ is introduced in computations where Q is the flow rate; this involves no extra work because a simultaneous measurement of Q must be carried out anyway if **representative measurement of the flow per se is desired,** and not just of its variations. Further, an on-site calibration of the equipment must be done.

3.2.2 Example of calibration (from Grange [17])

The sketch in Fig. 3.11 shows the assembly of equipment installed on the storm weir located downstream of a catchment area of 1065 ha of the agglomeration of Orlean, with combined drainage. The system is built around the pollutometer HORIBA (see Sec. 3.2.3) and is activated by an overflow which automatically triggers the process of measurement as soon as overflow starts at Loire during the rainy period. The sampler ISCO, similar to the one described in Sec. 3.1, is used for automatic drawing of samples needed for calibration of the pollutometer and determination of certain parameters whose continuous measurement is not possible. The perfect synchronisation of operations is ensured by a device of numerical control.

Calibration was done using 70 pairs of values over a large interval. A very satisfactory linear correlation was found between COD and UV on the one hand, and between SS and visible particles on the other (coefficient of correlation for the two cases respectively 0.94 and 0.82).

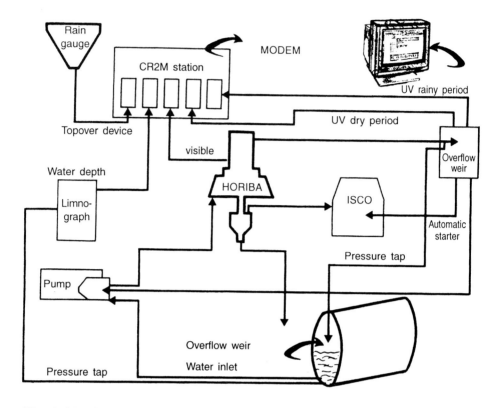

Fig. 3.11: *System of measurements, acquisition and teletransmission of data.*

The raw data from the pollutometer were first transformed to equivalent COD and SS by using the appropriate equations and then combined with the **synchronous** measurements of flow rates to obtain the file 'FLUX' at intervals of three minutes. In this way 23 overflows were analysed (the heaviest rainfall had a periodicity of about 20 days).

3.2.3 Available facilities

Some facilities are described below (selected to demonstrate advantages and drawbacks).

• **Horiba pollutometer** OPSA (Fig. 3.12, left)

Measurement of absorption in UV at 254 nm and visible at 546 nm
Advantages: Measurements possible at two wavelengths, one corresponding specifically to dissolved organic matter and the other to suspended matter; automatic cleaning and correction of optic derivative by eccentric rotation of the measuring cells (optic trajectory of variable length).

58

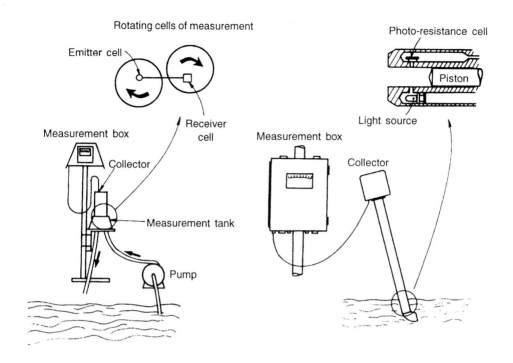

Fig. 3.12: Left: Horiba pollutometer; right: Monitek turbidimeter.

Drawbacks: Operation in derivation; hence an eventual absence of representativity of the sample under analysis if the pumping circuit is not properly designed; price rather high.

• **Ades pollutometer**

This pollutometer is marketed by Aquarius and was designed by the Open University of Brussels. **Rabier** (case study no. 4) has given a detailed description. It furnishes data of direct measurements [temperature, conductivity, water depth, turbidity (637 mm) and derived data (total SS and MO, identifiable and dissolved)].

One such apparatus is installed at Créteil and detailed information on initial calibration and intercalibration is available from this organisation.

• **Monitek turbidimeter** 53 LE (Fig. 3.12, right)

Measurement of turbidity in white light (400 to 700 nm)
The main advantages are direct immersion of the collector in the medium of measurement and its self-cleaning (the scraping piston picking up a sample every 15 seconds).

Drawbacks: The head of the collector is not submersible; hence a specific length is needed in the case of a major 'marinade' and its cost is quite high.

3.3 PROBLEMS ASSOCIATED WITH USE OF THIS EQUIPMENT

• **With the best instruments,** such as Horiba and Ades, it is possible to obtain very accurate measurements but they are expensive (about 250,000 F plus cost of installation, besides well-developed technical backup as well as provision for breakdown maintenance; thus the high initial cost). For this reason (for the time being), they should only be used for measurements of control or of short duration. A situation for their appropriate use is described by **Rabier** (case study no. 4); three Ades instruments were installed in the basin of Vitry-on-Seine to measure the efficiency of remediation of rainstorm run-off .

• **Soulier** (case study no. 6) describes another measurement device, still in the experimental stage, installed at Boulogne for monitoring discharge in the Seine River.

• **The simplest, though less accurate,** instruments may be used on large weirs for measuring the development of turbidity of flowing water by operating the discharge valves. As pointed out by **Breuil** (case study no. 9), this method is used in Seine Saint Denis at 23 surveillance stations. The problem of self-cleaning has still to be solved.

4. DATA NEEDED FOR MANAGEMENT

4.1 BASIC DATA

As mentioned in Chapter 1, Sec. 3, fundamental data—unfortunately, rarely available—for a network constitute a comparison between DWF from a treatment device and the discharge from the group of storm weirs (including that from the device if a weir exists there) during the rainy period (RWF) for an average year and for a storm event of a given periodicity (including that from the flow from separate systems).

The underlying assumption is that discharge from weirs (and rain) can be evaluated quantitatively as well as qualitatively; hence the need to:
— classify the overflow in increasing order of quantity so as to restrict the **evaluation and measurement to a limited number of works;**
— establish (or use) data from limnograph on pre-selected weirs;
— evaluate waste flows from a few measurements taken to assist in determination of the range within which major parameters of pollution lie.

As a first step to accomplishing the above, all wiers of the network as well as their main characteristics must be known. A description of Parisian weirs is given by **Constant** (case study no. 2). A catalogue of a weir of the Parisian network is given as an example in Fig. 3.13.

A document should also be available for each collector listing its various weirs as well as the surface area and characteristics of various catchment areas (zones 1 - 4 - 5 - 8 of Fig. 3.14) as illustrated by Table 3.6 (Dégardin [44]).

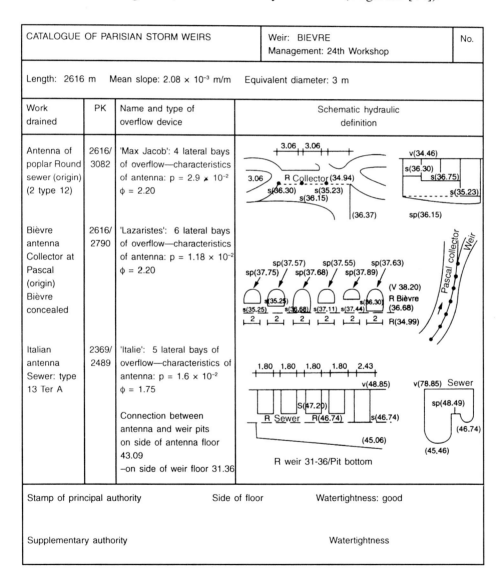

CATALOGUE OF PARISIAN STORM WEIRS			Weir: BIEVRE Management: 24th Workshop	No.

Length: 2616 m   Mean slope: 2.08 × 10⁻³ m/m   Equivalent diameter: 3 m

Work drained	PK	Name and type of overflow device	Schematic hydraulic definition
Antenna of poplar Round sewer (origin) (2 type 12)	2616/ 3082	'Max Jacob': 4 lateral bays of overflow—characteristics of antenna: $p = 2.9 \times 10^{-2}$ $\phi = 2.20$	
Bièvre antenna Collector at Pascal (origin) Bièvre concealed	2616/ 2790	'Lazaristes': 6 lateral bays of overflow—characteristics of antenna: $p = 1.18 \times 10^{-2}$ $\phi = 2.20$	
Italian antenna Sewer: type 13 Ter A	2369/ 2489	'Italie': 5 lateral bays of overflow—characteristics of antenna: $p = 1.6 \times 10^{-2}$ $\phi = 1.75$ Connection between antenna and weir pits on side of antenna floor 43.09 —on side of weir floor 31.36	

Stamp of principal authority	Side of floor	Watertightness: good
Supplementary authority		Watertightness

Fig. 3.13: Brief description of weirs of Bièvre.

Catchment area	Area in ha	Rate of impermeabilisation	Slope in ‰	Maximum length
1	333	0.65	0.46	5340
4	56	0.75	2.10	900
5	96	0.78	0.50	200
8	86	0.67	0.50	2000
Total	571			

Table 3.6: *Characteristics of catchment area of lower Bièvre*

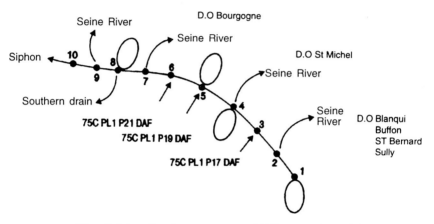

Fig. 3.14: *Catchment area of Bièvre weirs of Paris.*

In Fig. 3.14, the arrows at points 3-5 and 6 indicate the eventual contribution from the Bièvre collector, situated at a slightly higher level, which acts as a meshwork between the higher and lower Bièvre collectors. **Valiron** (case study no. 5) gives a detailed description of the zones of collection at Paris (Section 3) which readers might find useful.

The size of the bypass is important. The collector 'high Bièvre' has a catchment area of more than 8500 ha, of which 30 to 40% may be used by the 'low Bièvre', thereby bringing in a supplementary inflow of at least 60% under certain circumstances.

4.2 COMPLEMENTARY DATA NEEDED FOR STUDYING IMPROVEMENT OF DISCHARGE

For weirs whose estimated discharge is large and exceeds a certain value, it is useful to study the **discharge for various rainfalls in the catchment zone.** This study may be carried out by using the available models (see Chapter 5). The measures of control of flow rate and quality necessary for calibration of the model may also be useful, especially for computing the capacity of an eventual storage/decantation basin (see Chapter 4). For this it ought to be possible to add to measurements over the weir, the contribution of corresponding rainfalls. **Valiron** (case study no. 5) has given some data concerning the Bièvre.

It may be noted that in a number of cases, the data (stored in the software, sometimes at the central station) concerning hydraulic operation of modernised weirs can be used **but due to lack of human resources and motivation, actual use has been very limited!**

Maintenance of these plants and the associated facilities for measurement requires sufficiently strong teams of well-trained personnel. This aspect has been neglected. **To repeat, good instrumentation and teams of well-trained personnel are absolutely mandatory.**

Measurements of a series of actual overflows on the weirs are far more accurate than those of a rainfall used for calculation of planned works. **The difference is more significant when the catchment area is large and the network complex.** For the weir at Clichy, Serre [80] compared the decennial overflow calculated from a typical rainfall with a series of overflows measured during the 10-year period and observed that the difference was notable; the actual volume of discharge for one year was one million m^3 versus the calculated ten-yearly overflow of 650,000 m^3 (the actual would be about three times). For the weir of Briche whose catchment area is smaller, this difference was less, 232,000 m^3 for the decennial; the volume trebled in 6 months during 1993 and is still very large today (a factor of about 2.5 for a 10-year period).

4.3 EQUIPMENT FOR DYNAMIC MANAGEMENT

As suggested by **Delattre** [18], and to be seen in Chapter 7, the trend is gradually changing from **staid management** for which rules have been fixed once and for all, to **progressive management** for which, based on experience gained the rules are modified after an event (subsequent to proper analysis and understanding), to **dynamic management** for which the rules may be modified even during the event itself.

Progressive management requires a posteriori analysis of events and is largely facilitated if the following information is stored in the computer: progress of the event, changes in parameters indicative of the instantaneous situation, especially the location of major weirs, and volume discharged during drop in rainfall as measured by rain charts.

A more thorough analysis is possible if the flow of contaminants and their impact on the natural environment can be estimated.

The aforesaid need is well recognised and many managers have installed a number of measuring devices with direct input in their computer system. For

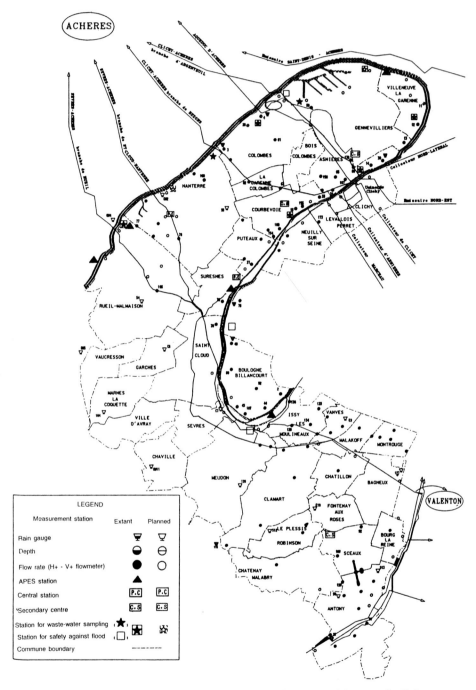

Fig. 3.15: *Measurement network of the district of Hauts de Seine.*

example, the network feeding the first trench of Valenton comprises 52 management sites. The departmental network of Hauts de Seine (Fig. 3.15) feeds its 50 control stations through more than 350 terminals and this number is likely to double in future, and comprise 23 rain charts and radar images of Trappes. Further, this system will be connected in real time with Calysto which manages the data on quality of the Seine River collected from three APES measurement stations situated in the district. **Obviously, storing such data in the computer makes sense only if it will be used!**

Dynamic management requires, in addition, 'continuous data' from the limnographs of major weirs and eventually from the turbidimeters and rain charts transmitted to the central station; it also requires a liaison of command with mechanisms for activating the valves of these works.

Attention of the user of a network managed in a dynamic manner by means of automatic devices, should be drawn to the response time of these works and their unforeseeable suddenness. **The safety of personnel involved with maintenance requires that all rules laid down by the district of Seine Saint Denis** (ref. h) be strictly followed. This task is facilitated if special provisions are made at the stage of design of the automatic devices, for example, the possibility of locking some equipment during inspection hours, which would strongly reduce their speed of action (Fig. 3.16).

In conclusion, it may be noted that dynamic management is possible only if the battle against inundation is kept in check and one is equipped with the means to handle the increased complexity of the network and the enormous mountain of data to be handled.

Fig. 3.16: *Safety precautions to be taken for automatic management of networks.*

CHAPTER 4

Evaluation and Improvement of Hydraulic Efficiency and Separation of Contaminants

1. EFFICIENCY OF WEIRS AND THEIR MEASUREMENT

1.1 HYDRAULIC EFFICIENCY

Figure 4.1 shows the hydrograph on a weir for a rainfall P_i from t_1 to t_2. Volume V_a is diverted downstream during the storm comprising three sequences: V_{a1} from t_1 to t_2 before overflow for the discharge Q_d from the weir, V_{a2} from t_3 to t_4 flows towards the plant during the overflow and V_{a3} from t_4 to t_2 is the volume V corresponding to the storm.

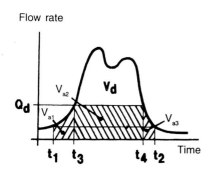

Fig. 4.1.

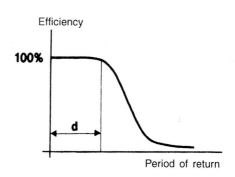

Fig. 4.2.

The hydraulic efficiency for this rainfall P_i is

$$E_{pi} = \frac{V_a}{V_a + V_d} = \frac{\text{Volume diverted to the plant}}{\text{Volume received during the storm}} \qquad (4.1)$$

The hydraulic efficiency is 100% for a light rainfall; or for a rainfall corresponding to the flow rate of Q_d of periodicity d; the efficiency decreases as the period increases, as shown in Fig. 4.2.

The efficiency for a year x is given by the following formula

$$E_{year\ x} = \frac{\sum_{1}^{n} V_{ai}}{\sum_{1}^{n}(V_{ai} + V_{di})}$$ (4.2)

In the numerator the summation is carried over the volume diverted downstream and the denominator represents the sum of total volumes of all storm events 1 to n resulting in rain-water run-off. Of these n events, arranged in the order of increasing total volume $(V_a + V_d)$, N events cause overflow.

These n rainfalls of total volume $(V_{ai} + V_{di})$, corresponding to the i-th event, are shown in Fig. 4.3. The broken line represents the volume diverted downstream.

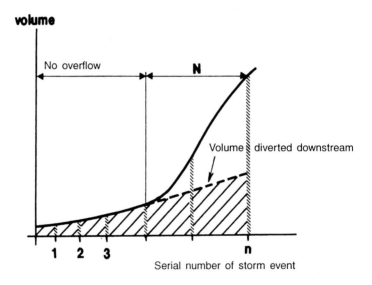

Fig. 4.3: Classification of rainfalls of the year x.

Obviously, $N = n - j$. The efficiency for the year is the ratio of hatched area and the area under the solid line curve.

The volume diverted downstream varies slightly due to a drop in efficiency when $V_a + V_d$ increases. The total duration T_x of storm events during the year is the sum of durations $(t_{2i} - t_{1i})$, corresponding to event n_i.

$$T_x = \sum_{1}^{n}(t_{2i} + t_{1i})$$ (4.3)

Variation in T_x from one year to another is much smaller than $\sum_1^n (V_a +$
$V_d)_{xi}$ because an exceptionally heavy rainfall can contribute in just a few hours 35 to 50% of the year's average volume. The efficiency over the year is smaller than

$$\frac{Q_d \sum_1^n (t_{2i} - t_{1i})}{\sum_1^n (V_{ai} + V_{di})} \tag{4.4}$$

The quantities for which **knowledge is essential in managing a sewerage network and its treatment plant** are volume of inflow to the plant during the rainy season and the overflows upstream to the natural environment. **The efficiency of the network is defined as the ratio of the first volume to the second.**

1.2 Efficiency of separation of contaminants in flow

It has been known for about twenty years that some weirs separate part of the contaminants, primarily the large particles, flotsam and part of the smaller suspended particles (according to **Green** [20], SS particles of $d \geq 0.5$ mm).

This separation by various types of weirs and their accessories is schematically described in Table 4.1.

Two criteria are defined for comparing the effects of one type of reserovir with those of another: **total efficiency and the treatment factor.**

With the help of the following formulas, the two criteria for a weir can be computed:

–total efficiency $\quad \eta = \dfrac{\Sigma P_b}{\Sigma P_i} = \dfrac{\Sigma Q_b \cdot C_b}{\Sigma Q_i \cdot C_i}$ \hfill (5.5)

–treatment factor $\quad \varepsilon = \dfrac{(\Sigma P_b / \Sigma P_i)}{\Sigma Q_b / \Sigma Q_i}$ \hfill (5.6)

where C is the concentration of contaminant, Q the flow rate and P the mass flow rate ($P = CQ$).

The indices i and b refer respectively to the inlet of the weir and the flux towards the plant. In conformity with the criteria defined above, the larger the value of η, the better the weir. As for ε, it is equal to 1 if the weir action is limited to dividing the flow (the concentrations upstream and downstream are equal). A value greater than 1 signifies that the effluent diverted towards the plant

Type of reservoir	Type of pollutants			
	Soluble	Finely suspended	Other solid particles	
			Floating	*Suspended*
Opening in wall	No effect	No effect	No effect	No effect
Opening in floor	"	"	"	"
Low weir	"	"	"	Traps part of low flux of solids
Low weir on side with skimming at surface	"	"	Small effect	"
Low weir on side with screen	"	"	Not as good as the one with skimming at surface	
High weir on side	"	"	No effect	Traps part of low flux of solids
High weir on side with skimmer	"	"	Highly effective at low flow rate Risk of resuspension at high flow rate	
High weir on side with screen	"	"	Screen reduces performance	

Table 4.1: *Effect of different types of reservoirs on the flux of contaminants*

is comparatively more polluted than that discharged in the natural environment. The larger the value of ε, the better the weir.

Measurements should be made during the storm episode, i.e., between t_1 and t_2 (beginning and end of the episode considered) of Fig. 4.1.

These coefficients denote the ratio of the pollution retained in the network to that which flows out.

With respect to the receiving environment, the relevant quantity is the difference of the mass that overflows and hence the ratio

$$1 - \eta = \frac{\Sigma P_i - \Sigma P_b}{\Sigma P_i}$$

The smaller the value of $1 - \eta$, the better the weir

1.3 MEASUREMENT OF EFFICIENCY

1.3.1 HYDRAULIC EFFICIENCY

For a storm event, computation of the hydraulic efficiency requires measurement of the volume discharged and that of inflow during the storm (see Fig. 4.10). Calibration of the weir and knowledge of the law $Q = f(h)$ help in computing $V_a + V_d$,

$$V_a + V_d = \int_{t_1}^{t_2} f(h)dt$$

Since the downstream flow rate Q_a is related to h by a law f(h) (see Chapter 3, Sec. 1.2), the volume V_a which flows towards the plant can be readily calculated in the case of a high weir as well as that of a constriction.

Computation in the case of a low weir is more complex because the flow becomes turbulent during some storm events; however, measurement of inflow at the upstream can be utilised for some large weirs.

Computation of the efficiency of the network for a storm event is facilitated by measuring the flow rate at the inlet to the plant, which gives V_a, while V_d is the sum of the volumes discharged at the weirs of the network. Alternatively, the total volume of flow over the zone can be calculated to obtain an approximate value of $V_a + V_d$ (see Chapter 1, Sec. 3).

1.3.2 Efficiency of separation of contaminants

This concept has only recently been introduced. Calculation of this efficiency is more difficult because the efficiency for the particles susceptible to separation has to be calculated individually and further, a distinction has to be made between floating and suspended particles. Furthermore, on-site measurements are a complex and expensive proposition.

• **Problem of flotsam:** This has been studied, notably in England, both from on-site and laboratory measurements (**Green** [20]). With a **Gross Solids Sampler** (GSS) it is possible to sample a portion of the flow by making it pass through a sac (Coposac) with perforations of 5 mm. The samples are taken alternatively from the flow at the inlet and the overflow (with a time delay to preclude contamination of the samples). A description of the sampler is given by **Walsh** [21] and **Jeffries** [22] who emphasise the correct position of the sampling tube (it is often necessary to provide a small reservoir for the sample taken from the overflow to ensure that it does not contain too much air). The results of measurements made at Ecosse are given by **Walsh** and **Jeffries** [23].

Another apparatus (Fig. 4.4), **Gross Solids Monitor** (GSM), works on a different principle. When the weir begins to overflow, the instrument pumps a

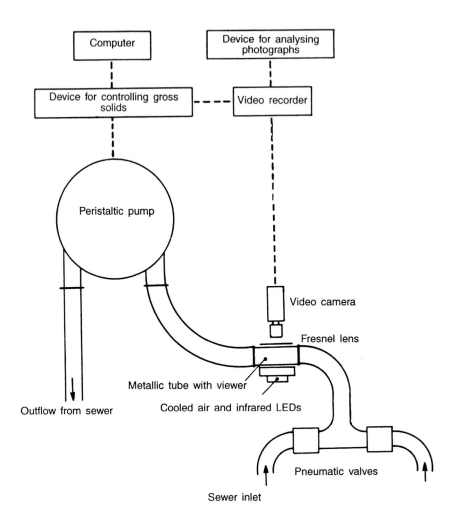

Fig. 4.4: *Gross solids monitor.*

large volume of water to the inspection chamber. By means of a flexible tube (**Cootes** [24]) the lower and upper portions of this chamber become transparent when illuminated from below by infrared rays. The original idea was automatic tracking of the shadows of the gross solids when they cross the chamber, enabling a count of these particles. But the idea did not prove practical and a video camera was installed to record the images and analysis was carried out away from the site. The system of automatic detection was never perfected; analysis was done instead by observers who studied the video and counted and estimated the size of the solids on the screen. Although the procedure was rather tedious, the results reported were good. In the case of on-site counting, it was found that the number of particles at the inlet varied as the square of the flow rate.

On another site, the sampler and monitor were placed in the same box and a float passed first through the monitor and then through the Coposac of the sampler, thereby joining the two instruments.

The studies carried out at Ecosse over a period of six months with a GSS yielded data for 22 events. The quantity of collected matter was calculated by multiplying the flow rate and measured concentration; these results were then compared with those from laboratory tests. **Jeffries** [22] deduced the following formula which he claims is valid for data at the inlet and on the weir and correlates the mass of gross solids and that of SS:

mass collected during an event $= 0.005e^{2\log}n^c$

with c the mass measured in the laboratory.

• Suspended matter

The cost of measurements and their complexity for obtaining a good representation prompted laboratory measurements for comparison of performance of various works: the settling chamber, high and low weirs (lateral, simple or double), vortex with peripheral weir (Fig. 2.13, Chapter 2) and the Storm King™ (see **Ruff** [25]). The reduced models were 10 to 20%—up to 50%—the normal size. The study was carried out on the same configuration (and storage capacity) and steady flow conditions with plastic or nylon granules with the same velocity of fall as the various components of SS.

Table 4.2, taken from **Hedges** [27], gives an idea of the relative efficiency of some weirs.

Results obtained in Germany and the USA by **Pisano** and **Brombach** [28] were similar but are quite old and were obtained with different granules.

Hedges [26] obtained a good correlation between laboratory and on-site measurements for the Storm King™ (Fig. 4.5).

	Lateral weir with double overflow	Frontal high weir with settling chamber	Weir with vortex effect	Storm King
Flow rate at inlet equal to nominal flow rate	$\eta = 15.2$ $\varepsilon = 1.02$	$\eta = 15.6$ $\varepsilon = 1.13$	$\eta = 17.0$ $\varepsilon = 1.13$	$\eta = 18.5$ $\varepsilon = 1.23$
Flow rate at inlet equal to one-half nominal flow rate	$\eta = 16.1$ $\varepsilon = 1.07$	$\eta = 16.7$ $\varepsilon = 1.11$	$\eta = 20.8$ $\varepsilon = 1.39$	$\eta = 29.1$ $\varepsilon = 1.94$

Table 4.2

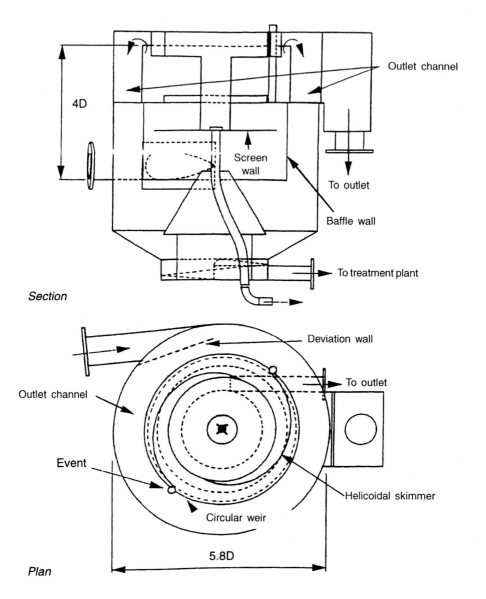

Fig. 4.5: *Hydrodynamic separator Storm King^TM.*

According to **Chocat** [14], there are two reasons for exercising caution during interpretation of results:

- they were obtained by different researchers and the methodologies used were not necessarily compatible;
- they were often obtained with constant flow rate at the inlet; steady flow conditions differ markedly from the actual conditions of a sewerage network.

Marchandise [29] evaluated the efficiency of the SST of Lievain (similar to that shown in Fig. 2.13 of Chapter 2) installed at the extremity of a sewer of 1500 Ø on a portion—800 ha (500 rural and 300 urban) of 4800 ha—of the catchment area. This SST, designed for a flow rate of 0.9 m³/s, comprises a cylindrical tank of 7 m dia; it supposedly provides a storage of the order of 400 to 450 m³.

The efficiencies of discharge of the SST and a standard high weir for the 27 storm events measured over a period of 5 months (November, 1986 to March, 1987) are compared in Table 4.3. The maximum flow rate was 792 l/s.

	Efficiency of SST in overflow	Efficiency of SST for all events	Efficiency of DO in overflow	Efficiency of DO in all events
Min	0.34	0.37	0.26	0.32
Max.	0.82	0.93	0.64	0.90
Average	0.56	0.69	0.44	0.61
Ec. type	0.14	0.15	0.13	0.16

Table 4.3: Efficiency of discharge of SST and DO

The SST measuring device is depicted in Fig. 4.6.

These measurements seem to confirm that the efficiency of the SST is significantly higher than that of a high weir (see Hedges, Table 2).

They also show that storage upstream, together with decantation, improves the efficiency of the work by as much as 69%.

Lastly, given the presence of screens and a partitioning siphon, the macrowastes and floating particles are not released into the river.

The results obtained on this SST by Marchandise are not very good but nonetheless better than the mediocre data obtained by the **DEA of Seine Saint Denis** [81] for the first work of this type installed in France. The difference in results can apparently be attributed to the difference in calibration of the SST, which made use of the storage in the upstream conduit. The SST at Lievain combines the effects of high weir and the vortex.

2. POSSIBLE IMPROVEMENTS IN WEIRS

According to **Crabtree** [31], for a weir to be efficient it must fulfil several conditions summarised in Table 4.4. He has also emphasised that automatic weirs

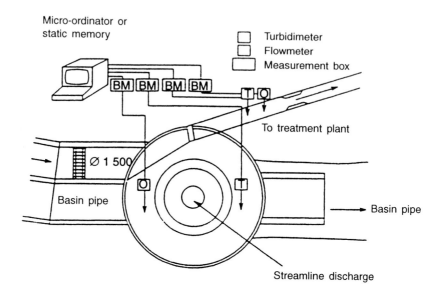

Fig. 4.6: *Installation of SST turbidity and flow probes (schematic).*

Hydraulics	—no discharge when the flow rate is less than the nominal for overflow —no increase in downstream flow rate during overflow
Pollution control	—floating matter not to be discharged —maximum diversion of floating matter downstream
Management	—should be simple and not require skilled personnel —should require little maintenance —should be self-cleaning —should have easy access

Table 4.4

are preferable and that only the four types of works listed below fulfil the conditions set forth in Table 4.4:

— weirs with high-level setting for overflow,
— lateral weir with settling chamber,
— installation with vortex,
— hydrodynamic separator.

The choice among these four types depends on the site and cost but according to **Ellis** [30], more than 150 Storm Kings have been installed during the last 5 years. Discussion of the possible choices will be resumed in Chapter 5, in particular the conditions of automation for static and adjustable works, since the latter are generally large networks requiring skilled management. Economic considerations based on cost data are also discussed. But only in terms of hydraulic conditions and those of discharge of contaminants.

2.1 EFFECT OF STORAGE UPSTREAM

Discretisation of flow from a storm event creates a peak flow rate and augments its duration.

Figure 4.7 shows the effect of linear storage when the discharge is adjusted at Q_s [by means of a regulator DR (Fig. 4.8a)] for a storm event contributing a volume V_1 which is smaller than V, the storage volume available in the basin. Discharge is shown by the dotted curve: $V_2 = V_1$. A basin of this type should be equipped with a weir for the case wherein the storm volume V_1 is larger than V.

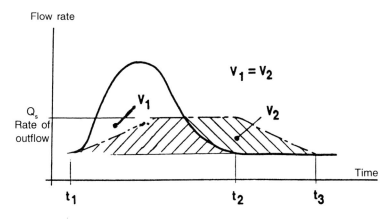

Fig. 4.7: Effect of storage on a storm event.

This weir can be replaced by a basin fed parallely (Fig. 4.8b) by means of a regulator DR_1. In this case, when the storm volume V_1 is larger than the storage capacity V, the flow rate downstream may exceed Q_s (Fig. 4.9) because of the drainage flow but V_2 is always equal to V_1.

The maximum flow rate is reduced to $Q_{lim} < Q_{max}$ and, in addition, the effect of storage is to increase the duration of passage of the event from $(t_2 - t_1)$ to $(t_3 - t_1)$, which increases as Q_s decreases.

It may be noted that the parallel installation has an advantage in that the dry-weather flow is carried by the conduit; however the regulator DR_1 is more complex.

77

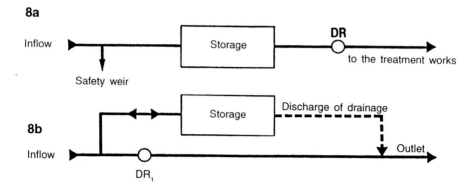

Fig. 4.8: Linear (8a) and parallel (8b) storage.

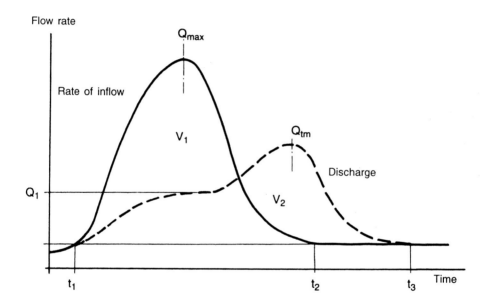

Fig. 4.9: Effect of parallel storage when $V < V_1$.

With a parallel installation it is possible to take advantage of a much larger latitude with respect to storage volume (possibility of evacuation by pumping) and management of evacuation (which can be varied over time).

It may be noted that to reduce the volume to be stored (of course, involving risk of aggravation of pollution downstream), evacuation should be as rapid as possible taking into account the flow rate permissible downstream. This means the response curve of the regulator should be almost vertical.

The major effect of these works is to discretise the flow, keeping two objectives in mind:

  – limit the size of the sewers downstream,
  – reduce the number of overflows from weirs situated downstream and increase the volume transferred to the downstream (reduction of Q_{max} and increase in passage time T_x).

Figure 4.10, taken from **Agg** [33], illustrates this point with the help of a chart used in The Netherlands. It gives the annual frequency of overflow as a function of the storage capacity (including that of the network) and the value of the

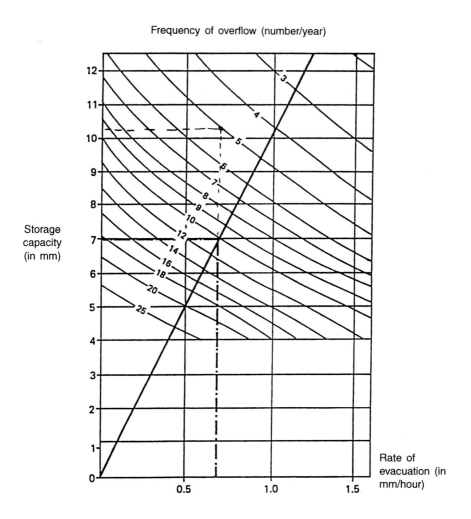

Fig. 4.10: Kuiper's relation between storage capacity, evacuation rate and frequency of overflow.

discharge (the storage of the network is the volume contained between the surface of dry-weather flow and that of the surface corresponding to overflow).

The thick line in Fig. 4.10 represents the time of evacuation equal to 10 hours in the absence of a fresh rainfall. The millimetres are relative to the impermeabilised surface. The example shows that corresponding to 0.7 rate of evacuation and 10 overflows/year, the total storage capacity is 7 mm. Taking into account the capacity of the network, which is large in The Netherlands—a flat country—this corresponds to detention in a basin of the order of 2 mm, or 20 m³/ha when the frequency of overflow is 5/year (**Voorhoeve** [34]).

The results of a study carried out in the Netherlands on discrete detention works are given in **NWRW** [35].

Today, given the tools of simulation of networks, it is possible to integrate the hydraulic effect of these discrete detention tanks and to measure the effect on the weirs; it is thus possible to **select the most favourable volumes and locations.**

These detention tanks can be located just at the upstream of the weir and can be constructed **either by enlarging the section of the sewer or by drilling a tunnel parallel to it at such a level that evacuation by gravity or pumping is possible. Underground tanks can also serve as storage tanks.** The other arrangements required for these works are expensive (**Valiron,** ref. A). They consist of **personnel safety** (ventilation and gas detector), **maintenance of cleanliness,** devices for **measurement of level** and command of evacuation connected to a management station.

In addition to their hydraulic role, these works contribute indirectly to the pollution discharged in the natural environment (or transported to the plant).

Their function of retarding the flow of contaminants downstream as a result of storage of part of the flux, causes the particles to be deposited and the quantity deposited varies indirectly with the time of water retention (or degree to which the flow is slowed down).

This secondary effect of decantation varies with the type of work:
- It is small if the sewer is overdimensioned; nevertheless the quantity deposited increases and concomitantly the amount of flushing required.
- In the case of storage in a tunnel, the effect of decantation depends essentially on the time of detention; evacuation may cause resuspension of the deposited material.
- In the case of storage in a basin, the effect of decantation is related to the time of detention as well as the shape of the reservoir and the precautions taken during evacuation. The performance can be the same as that of basins of decantation for the same duration of detention.

According to **Bellefleur** [32], 155 storage basins with a total capacity of 37,000 m³ have been installed at Bas Rhin in the past 15 years.

The **settling chambers** described in Chapter 2, placed just upstream of the weir, play the same role.

Figure 4.11 shows the effect on the downstream flow rate of the flux stored between t_1 and t_3; the volume that overflows is reduced by an amount V_s above the overflow rate Q_d and the volume diverted to the plant is increased by an equal amount, given the increase $t'_2 - t_2$ during the passage time of the storm event. The efficiency of the weir (see formula (4.1), Sec. 1.1)

$$R = \frac{V_a}{V_a + V'_d} \quad \text{becomes} \quad \frac{V_s}{V_a + V'_d}$$

and is thus increased due to the stored volume V_s; the increment is equal to the ratio of Q_d to the total volume of the storm event.

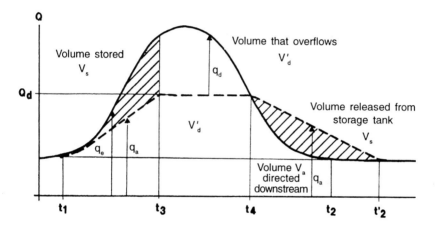

Fig. 4.11.

2.2 REDUCTION IN LOAD OF CONTAMINANTS DISCHARGED AND STORM BASINS

When the weirs, after incorporation of the improvements discussed above, fail to meet the conditions set forth in the regulations given in Chapter 6 with respect to the pollution brought in, the quantity of contaminating material discharged should be reduced by building basins as briefly described in Sec. 2.4 of Chapter 2.

2.2.1 Basin design

Figure 4.12, taken from **De Carne** [36] and **ATV** [37], shows two tanks constructed on a basin in which the time of detention is 10 minutes. The ensemble comprises a tank for the 'first flow' and three weirs plus an extra for safety, with a 'skimmer' for trapping floating matter—an arrangement widely used in England.

It was found that with these rectangular tanks, efficiency of treatment improved and cost of maintenance reduced when the following specifications were adhered to:

— depth of water between 1.6 and 2.5 metres;
— rectangular shape with breadth equal to 1/2 or 1/3 length;
— compartmentalisation into two tanks to facilitate cleaning;
— no water turbulence at the inlet.

These days, circular tanks are being constructed in which it is possible to install adjustable devices for cleaning, similar to the scrapers used in decanters designed by STEP.

Figures 4.13 and 4.14, taken from **Stahre** (ref. K) show the configuration of the bottom of several rectangular tanks to facilitate cleaning. Such tanks are widely used in Switzerland and Germany, as they enable diversion of deposited matter to the treatment plant.

Washing devices are often operated from a container, such as the one pictured in Fig. 4.15[1] .

Similar to the discrete detention tanks mentioned in the previous section, these tanks should also be provided with a gas detector, ventilation and passageway for persons and material needed for flushing[2]. In most such tanks, pumps are provided to divert the stored water to the network at the end of a storm.

Figure 4.16 shows the installation of pumping devices on a circular tank in Hamburg (Germany).

Figure 4.17 gives a sketch of a cylindrical tank with tangential inflow that facilitates both decantation and cleaning of the bottom.

These circular works are best inserted in inhabited areas, given their reduced surface area, and are preferred to the standard rectangular works. Many examples of construction of such works in Germany, Switzerland and eastern France are given by **Valiron** (ref. A) and **FNDAE** [38].

2.2.2 Computation of storage capacity and decantation

In Germany, such computation is based on the directive **ATV 128** [37], which has been implemented for many works. The recommended storage volume is such as is necessary to ensure that the pollution carried in the rain-water flow discharged in the environment (= pollution discharged in the D.O. + pollution discharged after treatment at the downstream of the STEP) does not exceed the

---

[1] A device of this type is also shown in colour plate no. 10.
[2] The cleaning operation in a tank by a motorised cart brought through a trap door is shown in colour plate no. 11.

Side view

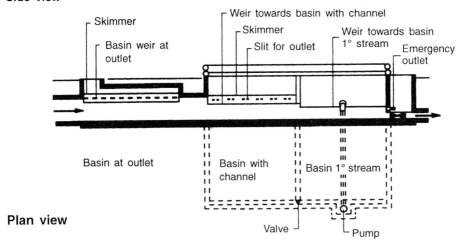

Plan view

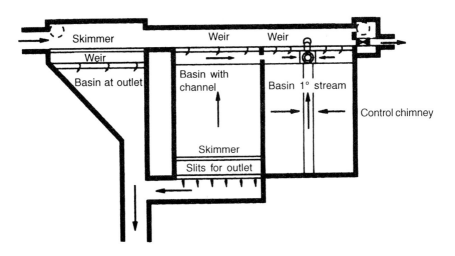

Fig. 4.12: A typical double tank for decantation used in Germany (from ATV [37]).

rainstorm pollution rejected on the hypothesis that the network functions as a discrete system.

Application of this method is illustrated in Fig. 4.18. One drawback is that the time of detention of the flux transiting in the basin is not fully taken into account. The volumes computed by this method are too small for good decontamination.

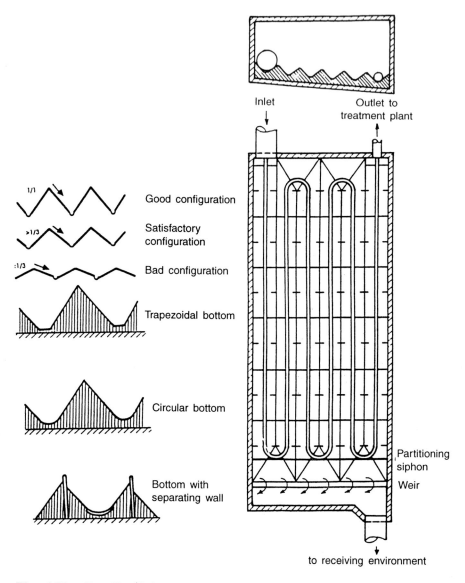

Inlet

Outlet to treatment plant

1/1 Good configuration

>1/3 Satisfactory configuration

:1/3 Bad configuration

Trapezoidal bottom

Circular bottom

Bottom with separating wall

Partitioning siphon

Weir

to receiving environment

Fig. 4.13: *Details of three types of gutters (from **Stahre**, ref. K).*

Fig. 4.14: *Tank with continuous gutter (from **Stahre**, ref. K).*

Another method which gives more accurate results is that of the American norm **EPA** [39]. This method, together with details of computation, is described in Annex. I (document 2). (This document as well as document 3, has been taken from **Valiron**, ref. A.)

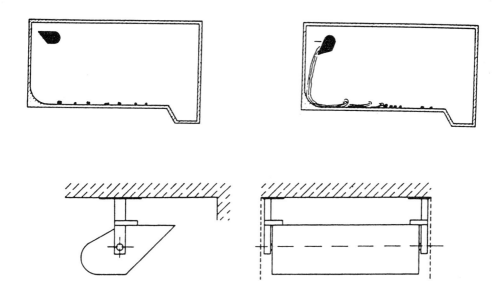

Fig. 4.15: *Washing-rinsing devices.*

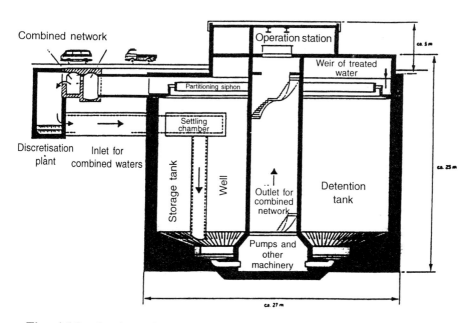

Fig. 4.16: *Section of a circular tank in Hamburg (Germany) (schematic).*

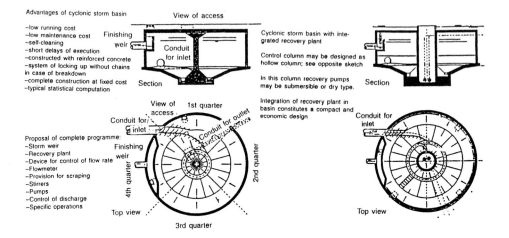

Advantages of cyclonic storm basin

–low running cost
–low maintenance cost
–self-cleaning
–short delays of execution
–constructed with reinforced concrete
–system of locking up without chains in case of breakdown
–complete construction at fixed cost
–typical statistical computation

View of access

Finishing weir

Conduit for inlet

Section

Cyclonic storm basin with integrated recovery plant

Control column may be designed as hollow column; see opposite sketch

In this column recovery pumps may be submersible or dry type.

Section

View of access

1st quarter

Conduit for inlet

Conduit for outlet

Finishing weir

4th quarter

2nd quarter

Top view

3rd quarter

Integration of recovery plant in basin constitutes a compact and economic design

Conduit for inlet

Top view

Proposal of complete programme:
–Storm weir
–Recovery plant
–Device for control of flow rate
–Flowmeter
–Provision for scraping
–Stirrers
–Pumps
–Control of discharge
–Specific operations

Fig. 4.17: *Cyclonic decantation basin designed by the District of Metz (Vogelsberger brand).*

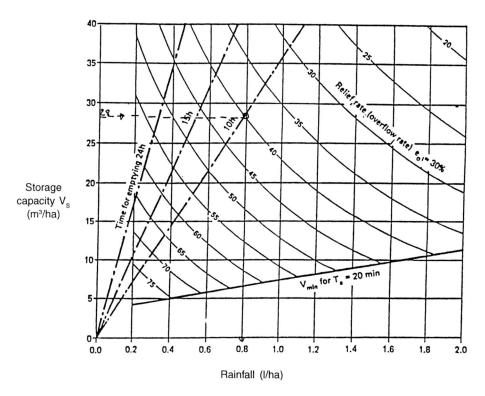

Fig. 4.18: *Chart for computation of storage volume (from ATV, ref. 3.2).*

Nowadays, simulation models of networks (see Chapter 5) are used for such computations; they are supplemented by a qualitative module for describing the actual situation and testing the impacts of the scenarios. Simulations are carried out either over specific episodes or periods of long duration.

2.2.3 Efficiency of basins

As elaborated in the norm EPA, **the efficiency depends on the retention time, particle grain size and pattern of water currents in the work.**

The retention time necessary for obtaining a specified percentage reduction of SS in quiescent water is shown in Fig. 4.19.

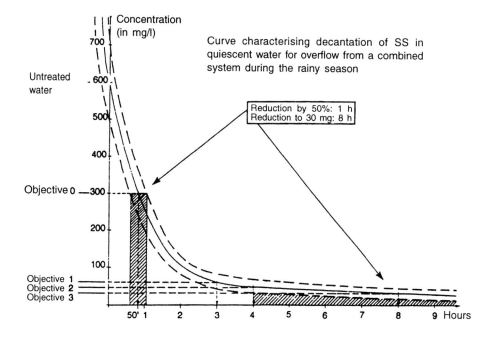

Fig. 4.19: Curves characterising the decantation of SS in quiescent water (Department of Seine Saint Denis, Directorate of Water and Sanitation).

The efficiency depends on the volume of trapped rain-water also. For light rain, less in volume than the capacity of the basin, the efficiency is 100%; with heavy rainfall, however, the efficiency drops rapidly. In fact, part of the contribution of the heaviest annual rainfall is not stored but flows away. **Chebbo** [4] has used the French data on the RUTP to compute the storage volume required for trapping 80% of the influent mass (Table 4.5).

		SS	COD	BOD
Volume required in m³ per impermeable ha	Min	33	33	33
	Medium	51	48	50
	Max	64	62	68

Table 4.5

Table 4.6 shows the storage volume required for trapping 80% of the influent mass during an event having the largest mass; increase with respect to the corresponding numbers in Table 4.5 is appreciable.

		SS	COD	BOD
Volume required in m³ per impermeable ha	Min	64	65	29
	Medium	88	107	88
	Max	120	134	144

Table 4.6

The data given in these two Tables indicate that the actual reduction of SS obtained in the overflow from basins hardly exceeds 60%. This relates to the overflow for rainfall of the periodicity for which computations were made assuming that the volume corresponds to 1 hour of the transit time; this volume increases very rapidly with the period of the cycle.

From the brief data on efficiency given in the chart of Fig. 4.18 or curves of the EPA norm, it is possible to determine approximately, for a storage volume relating to a weir, the efficiency of remediation for n different storm episodes studied during a year (see Fig. 4.3).

For the case of large discharge of some contaminants, it is considered advisable to provide a lamellar decanter and chemical treatment at the downstream of the storage-decantation basin; Fig. 4.20 illustrates the concept.

The lamellar decanter with flocculation capable of treating a flow rate Q_1 is set into operation about 30 minutes after the flow of water into the basin to allow time for activation of the flocculant. The discharge Q_1 slows down the process of basin filling, which can handle a maximum flow rate of $Q_{max} - Q_1$. The mass of contaminant added is thus reduced.

In this device the diversion, after rainfall, of water stored in the basin to the treatment plant is avoided; the treatment plant receives only the mud which is decanted and repumped after adequate dilution. If the situation permits, the mud can be trucked to the treatment site.

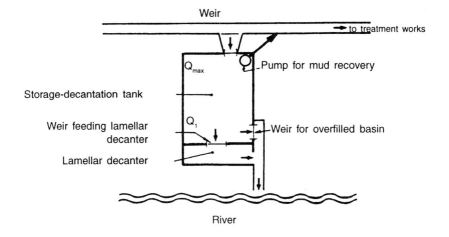

Fig. 4.20: Lamellar decantation on a storm basin.

Le Scoarnec (case study no. 7) modelled a solution of this type for the large weir of Clichy and found the most economic solution by assuming the stored m^3 costs 3000 francs and the treated m^3 costs 30 million francs. To suppress discharge of untreated, undecanted water during the period of study—March 92 to August 94—when the discharge was 35 million m^3, the most economic solution consisted of a combination of treatment and storage. This required a basin of 500,000 m^3 and lamellar decantation rate of 25 m^3/s (see **Le Scoarnec**, case study no. 7). A basin of this type with provision for lamellar decantation is under construction at Nancy (see Chapter 6, Sec. 3.6).

Valiron (case study no. 5) has selected 20 sites for storage-decantation basins to be constructed by the SIAAP. A capacity of the order of a million m^3 is required for a reduction by 70% in the pollution discharged during rainfall of a one-year cycle. The SIAAP is currently studying an alternative solution—to replace the basins which are difficult to construct with storage galleries playing the additional role of meshwork from one sewer to another.

CHAPTER 5

Placement, Calibration and Choice of Weir and Its Improvements

1. PLACEMENT AND CALIBRATION OF WEIRS

Placement of weirs is decided through a compromise made at the time of construction of a combined network, considering the four constraints listed in Table 5.1

Constraints	
Physical	Topography: slope, hydrographic basins, existence of natural water drainage etc. Land use: density and activities of inhabitants, roadways etc.
Environmental	Protection of natural environment Protection of inhabitants against foul odours, noise etc.
Economic	Cost of sewers vis-à-vis that of weir and related facilities
Management	Mode of management: static, dynamic Operational facilities: access, cleaning, maintenance etc.

Table 5.1

Some of these constraints are incontrovertible, e.g., the topography, while others change with urbanisation, i.e., land use and priority assigned to the environment, or with technological progress in terms of cost of works and modes of management.

All of the listed constraints are specific to each network, although with experience one does work out general guidelines as a basis for improvements and modifications that will need to be carried out, giving due consideration to development of urbanisation and advancement in technology.

91

1.1 Topography

Topography determines the broad specifications of the network in each supply basin to the town and depends on the natural water drainage which persists in spite of urbanisation. Location or placement of the first weir upstream of a catchment area is determined by economic considerations: cost of the weir including its outlet and extension of size of the downstream main sewer. The latter, approximately downstream of the natural brook and the downstream sewer parallel to the river are shown on the left side of Fig. 5.1. Curve 1 of the flow rate into the sewer as a function of the distance 1 measured from upstream is shown on the right side of this Figure. The first weir D_1 is situated at a distance l_1 and the discharge Q_1 is too large for the downstream sewer (because the downstream slope reduces, necessitating thereby a large increment in diameter). The sewers of D_1 and D_2 receive a discharge varying between Q'_1 and Q_2 (curve 2); the rate of flow into the downstream sewer is limited to Q'_2 by weir D_2.

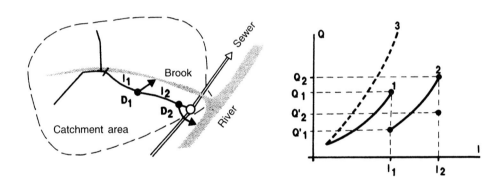

Fig. 5.1.

Curves 1 and 2 correspond to a particular state of urbanisation. With growth in urbanisation, the discharge reaching D_1 (curve 3, broken line) will increase and will necessitate:
— either doubling the size of the sewer,
— or installation of a discrete detention tank,
— or new construction that incorporates discretisation of discharge so as not to increase Q_1 and Q_2.

Figure 5.2 shows that in flat regions it may sometimes be necessary to provide weirs with overflow channels; but such channels can be expensive. This is the

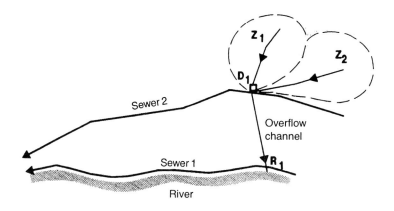

Fig. 5.2.

case of weir D_1 corresponding to catchment zone Z; the extension of the sewer between D_1 and R_1, in order to connect it to the main sewer 1, becomes more expensive than constructing an overflow channel D_1 R_1. (Such is the case for several Parisian weirs bringing water from the suburbs to a conduit flowing towards Achères, for example, for the outlet of Vincennes). Sometimes this outlet can be used for redirecting part of the storm run-off from the separately equipped zone Z_2 to the river.

1.2 CONSIDERATION OF URBAN DEVELOPMENT AND PRIORITIES

• Effects of urbanisation

A case in which urban development in basins 1 and 2 has necessitated construction of discrete detention basins r_1 and r_2 in order to avoid modification of weirs DO_1 and DO_2 is depicted in Fig. 5.3. But discretisation so achieved does not necessarily limit, in a satisfactory manner, the overflow to DO situated in the zones of the network where storm pollution involving large risks is produced. For example, in the case of Besançon, the excavated basins r_1 and r_2 situated on the steeply sloping low catchment area (small deposition in the sewers) help to reduce the volume of discharge on DO_1 and DO_2, but have only a limited effect on DO_3 which is situated downstream of a section of small slope characterised by a large sedimentation of flow during the dry season and thus discharges much larger quantities of contaminants than do DO_1 and DO_2 in the rainy season. A storm basin BO_3 had therefore to be constructed when priority was assigned to reduction of discharge to the river.

93

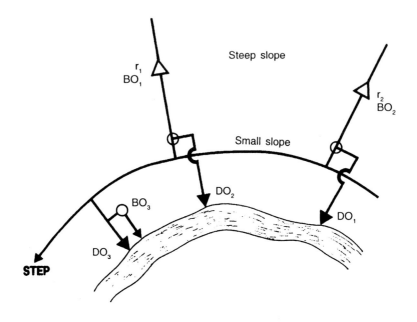

Fig. 5.3.

It is thus obvious that analysis of the effects of storm basins is rather complex and requires detailed analysis, aided these days by modelling of the network (Sec. 2.2).

- **Modernisation of management and equipment for weirs**

In large networks managed by competent personnel, the passage from static to progressive and then dynamic management (see Chapter 3, Sec. 4.3) leads to modernisation of weirs with maximum discharge, by equipping them with automatic valves and remote control devices. The cost of equipment needing civil engineering intervention can be reduced if the thresholds of some weirs are elongated, thereby enabling reduction in their number.

For example, the works at Hauts de Seine show that a net advantage can be expected from such an operation when combined with recalibration of works. The simulations carried out on 400 hectares of basins of Neuilly and Levallois and 1100 hectares of basins of Suresnes at Asnières, show that by raising the setting levels from 0.2 to 0.3 m and elongating the sills, together with provision of sector valves, it is possible to reduce the number of weirs by 50% and the discharge by 70%, given the fact of increased storage capacity thus achieved in the network without causing overflow.

Soulier (case study no. 6) has presented the studies in progress in the Boulogne loop, concerning measurement of the pollution discharged into the Seine River during the rainy season and designing the model MOUSETRAP to study this discharge.

According to **Ellis** [30], reducing the number of weirs is preferred in England.

It may be noted that the modifications to be incorporated in old weirs as a follow-up of municipal growth and new objectives are rather involved because they must take into account numerous factors which are difficult to isolate and must be integrated into costing.

A good approach is to choose the best solutions available, make use of the data on costs and proceed to simulations by means of modelling.

2. SELECTION OF IMPROVEMENTS IN WEIRS AND THEIR ENVIRONMENT

Table 5.2, taken from AGHTM [64], contains the various technical facilities available for achieving the three objectives that may help in fulfilling the new regulatory requirements.

Main objective	Technical means	Secondary effects
1. Reduce overflow and number of weirs	–Alternative techniques upstream of site –Storage-lamination basin –Overdimensioning of upstream sewer –Transformation to weir with high setting level	–Small reduction of influent pollution for some techniques –Effect of reduction of pollution depending on location of basin –Small effect on pollution
2. Reduce pollution discharged or take suitable action in the river	–Screen or skimmer –Weir with vortex or centrifuge –Storm basin –Lamellar decanter –Injection of air or oxygen for certain storms	(to eliminate gross solids) ←also reduces volume of discharge
3. Improve overall management of network	–Probes for measuring level, flow rate and pollution –Automatic valves and pumps or remote control	Reduces volume of discharge and pollution from weirs and storm basins

Table 5.2

To the list of techniques already described, in particular in Chapter 4, may be added the alternative techniques and actions given in the regulations for urbanisation relating to limiting the flow downstream, primarily in new urban centres, to the earlier accepted level[1]. A succint description is given by **Valiron** (ref. A) and a detailed description by **Azzout** (ref. G).

Injection of air in the river during a storm episode to counter oxygen consumption has also been tried; the procedure is detailed in Chapter 6, Sec. 3.6.

Table 5.2 highlights the fact that many techniques are complementary or concurrent. The choice depends on the 'cost effectiveness'of the technique. Unfortunately, this information is presently not available, even in terms of statistics based on analysis of existing data. **Such an analysis, when done, has shown large deviations depending on the conditions of execution and the constraints of the site (foundation, soil cluttering etc.) and even hampering of construction.** For example, due to lack of available terrain, the plans of construction of some storage-lamination basins had to be modified by increasing the diameter of the sewers—by a factor of two or three—or by providing a tunnel from which the water has to be repumped at the end of a storm event. A standard arrangement consists of a set of four pipes, three of which have the same diameter and the fourth—smaller in diameter—located at a lower level to carry the flow during the dry season or light rainfall. An entry chamber large enough for inspections and cleaning should comprise slots to enable closure of each pipe by movable lids[2].

At the downstream, the outlet chamber is provided with a device for regulating the flow rate and another device for isolating each pipe for cleaning purposes.

Figure 5.4 depicts the water line for the discharge determined by the capacity of the downstream structures and the position of safety devices placed either upstream (upper sketch) or downstream (lower sketch). If the storage basin is sufficiently close to an upstream weir, it can provide the required safety.

Thus these possibilities need to be examined, network by network, taking into account the local conditions. For classifying various solutions and eliminating some, one needs to consider only the order of magnitude of investment and operational costs; some available data are given in the next section.

2.1 SOME COST DATA

2.1.1 Alternative solutions

Table 5.3, taken from **Lyonnaise des Eaux** (ref. H), compares the investment (per m²) and maintenance cost (per m² /y) of a standard sewerage network for drainage areas of different sizes according to four alternative solutions.

---

[1] Coloured plate 12 pictures a permeable parking lot.
[2] See coloured plate 13.

Surface area of plot (in m²)	1000	900	800	700	600	500	400	300	200
–Traditional sewerage (canalisation) investment	7.2	7.7	8.3	9.0	9.9	10.9	12.1	13.5	14.6
–Cost of initiating operation (F/m²)	0.03	0.03	0.03	0.03	0.03	0.04	0.04	0.04	0.04
–Nidaplast wells investment (F/m²)	5.5	6.1	6.9	7.9	9.2	11.0	13.8	18.3	27.5
–Maintenance cost very small									
–Ballast roads investment (F/m²)	4.20	4.6	4.2	6.0	6.9	8.3	10.4	13.9	20.8
–Maintenance cost very small									
–Trench investment (F/m²)	5.0	5.4	5.7	6.1	6.6	7.2	7.8	8.3	
–Maintenance cost very small									
–Water meadow investment (F/m²)	2.5	2.6	2.9	3.1	3.5	3.9			
–Maintenance (F/m²/y)	0.6	0.6	0.6	0.7	0.7	0.8			

Table 5.3: Investment and maintenance costs of various sewerage solutions (1993 prices)

Valiron (ref. A) has given data on the cost of roadside reservoirs (Annex II) and a detailed quantitative comparison of various compensatory techniques.

An ever-increasing number of regulations with respect to sewerage place restrictions on these solutions and impose limits on the overland flow in new urban areas; they are a follow-up of regulations framed by the **District of Seine St. Denis** (ref. g) and the **Urban Community of Bordeaux** (ref. H).

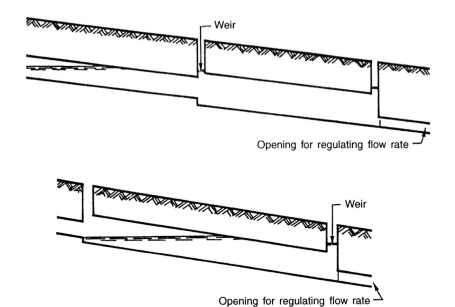

Fig. 5.4.

2.1.2 Storage-lamination basins

Table 5.4 (also taken from ref. H) gives some idea of effects obtained from reducing floods and reduction of pollution for storms of various return periods and frequency of overflow.

With respect to investment for underground tanks, the general rule of reduction of unit price (per m²) with the size is valid. It has been observed (see **Valiron, ref. A**) that the cost of works constructed in France is lower than that of similar works constructed in Germany or Switzerland, no doubt because the latter are more complex; they often include many tanks and very sophisticated washing and safety installations. The cost of equipment may be as high as 20 to 30% of the cost of the civil works. Another factor may be the more expensive foundations, which can raise the cost of civil works (80% of the total cost) by 30 to 40%.

Table 5.5 gives the available approximate data for underground tanks and storage in conduits or a tunnel.

Maintenance costs amount to at least 1% of the investment.

2.1.3 Cost of weirs[1]

This cost depends on the diameter of the main sewer and the length of the outlet and represents roughly the cost of 50 to 100 metres of the main sewer. To this

---

[1] 1993 prices

98

Function	Volume of basin relative to impermeable sector of zone			
	400-250 m³/ha	*250-150 m³/ha*	*150-75 m³/ha*	*75-30 m³/ha*
Flood reduction	Very good or good up to decennial frequency	Average to mediocre	Small to nil	Small to nil
Pollution reduction for exceptional storm episodes (period of return 2 to 10 y)	Very good to average	Average to mediocre	Mediocre	Nil
Pollution reduction for large storm episodes (2 to 5 y)	Very good	Good to average	Average to small	Small to nil
Annual pollution reduction	Very good	Very good	Good	Average to small
Frequency of overflow of partially treated water	1 to 10 times in 10 years	Twice per year to once in 5 years	2 to 10 times per year	10 to 30 times per year

Table 5.4: Efficiency of basins according to their volume and function

Storage-lamination basins	Germany-Switzerland: 7000 F (500 to 1000 m³) to 2500 F (50,000 m³) France: 5000 F.....to 1500 F
	+ 10 to 25% for equipment ⎫ of cost + 30 to 40% for special foundations ⎭ of civil works
Storage in conduits	2000 to 1500 F depending on diameter (2500 to 1500 mm)
Storage in tunnel	2000 to 3000 F depending on diameter and nature of the soil

Table 5.5: Cost per m³ of storage-lamination detention basin (1993 prices)

99

must be added the eventual settling chamber whose cost depends on the volume stored (5000 to 6000 F/m³), the valves and the outlets (their cost may be estimated as the same as that of the sewer).

Evidently the cost of renovation of a weir depends on the size of the works to be undertaken and the nature of the equipment. For example, the cost of renovation of weirs of Hauts de Seine, including the sector valves, varied between 5 and 20 million francs for an overflow of 2 to 10 m³/s. A significant part of the cost arose from the need to dewater during the work. Installation of an automated weir on an existing 'bay' of a weir with a sector valve in Ville de Paris cost 5/6 million francs. Costs for the Val de Marne are similar. It is assumed that the incoming flow has been diverted.

According to the **Guide des Vannes** (ref. B), the cost of valves represents 1 to 2% of the cost of works. A sector valve of 1 metre diameter costs about 400,000 F; its maintenance is around 10% per annum.

The cost of weirs with a vortex or centrifuge, the shape of which is more complex, is higher (by 20 to 40%).

2.1.4 Storm basins and associated lamellar decanters

• **These storm basins** resemble storage-lamination basins but are often more complex due to the presence of several weirs as well as better pumping facilities. Further, their installation seldom coincides with that of the associated weir and thus involves connecting conduits which may sometimes be long. This explains the extra cost of 20 to 30% over the data of Table 5.5.

Obviously, these costs are only indicative; unlike other works, an accurate evaluation here can only be done after a pre-project brief of the installation.

With respect to the efficiency of pollution remediation, data on SS given in Section 2.2.4 of Chapter 4 should be reviewed. Table 5.6, taken from **Lyonnaise des Eaux** (ref. H), reveals the importance of the effects of entrainment of other forms of pollution; the rate of attachment to SS based on experimental results obtained by **Bachoc** at Bordeaux [45] is given. But since these experiments pertain strictly to storm run-off, prudence should be exercised while looking at the results.

• **Lamellar decantation:** Tables 5.7 and 5.8, taken from **Lyonnaise des Eaux** (ref. H), show the efficiency, with and without chemical additive, measured at Bordeaux on a pilot with recirculation of mud on the water of Bequigneaux basin.

The results obtained with a simulated flocculant are good but it may be noted that a **work of this type must be associated with a basin** because, according to the results of tests with variable flow rate, it takes about 10 minutes to reach maximum efficiency.

The cost of a clariflocculator capable of treating 1 m³/s is of the order of 30 million francs.

Form of pollution	Concentration	Attachment to SS	Rate of elimination by decantation	Observations
SS COD BOD$_5$	248-464 mg/l 103-388 mg/l 35-97 mg/l	75 to 80%	79 to 87% 68 to 87% 73 to 87%	Results of experiments on holding basin at Bequigneaux (Bx)
Hydrocarbons	0.2 to 4 mg/l	42 to 84%	36 to 88%	–do–
Metals: Cadmium Lead Zinc Copper	3 to 60 mg/l (2) 50 to 1500 mg/l (2) 150 to 2000 mg/l (3) 30 to 1000 mg/l (2)	10 to 60% (1) 80 to 96% (1) 40 to 78% (1) 60 to 87% (1)	to be verified 82% (3) 67% (3) 72% (3)	(1) Measurements on experimental catchment area of Gradignan (Bordeaux) (2) Limits given in literature (strong variation depending on site) (3) Average figures from literature
Aromatic polycyclic hydrocarbons	200-3000 ng/l	Predominance on SS	To be verified	Attachment to particles (see Gradignan, Bordeaux)
PCB	Variable	90 to 93%	To be verified	Sometimes non-discernible (see Gradignan, Bordeaux)
Pesticides	Variable depending on type	Very small	Very small	
Pollution	High		Small	Often exceeds norms for bathing

Table 5.6: *Rate of attachment of various pollutants to SS*

Serial number	Lamellar velocity (m/h)	SS at inlet (mg/l) (average)	SS at outlet (mg/l) (average)	Mean efficiency (%)
1	12	208	90	56.6
2	21	266	130	51.1
3	25	293	92	64.3
4	30	281	142	49.4
5	35	564	236	58.2
6	47.6	200	155˙	22.4

Table 5.7: *Efficiency of SS removal without chemical additive*

Serial number	Lamellar velocity (m/h)	Dose FeCl (mg/m³)	Dose A32 (g/m³)	SS at inlet (mg/l)	SS at outlet (mg/l)	Mean efficiency (%)
1	30*	12	*	98	40	59
2	48	33	2	116.4	6.4	94.5
3	49	30	2	75.8	9.2	87.9
4	0→50	30	2	243	15.4	93.6
5	0→50	15	1.5	186	29.3	84.3
6	0→50	15	1.5	127	26.7	79
7	80	30	2	109.8	11.8	89.2

*Polymmer AS34—1 g/m³ (not AS32 for this test)

Table 5.8: *Efficiency of SS removal with chemical additive*

2.1.5 Devices for measurement and remote control

As seen in Chapter 4, knowledge of the network and weirs necessitates measures which help in choosing the exact location of the installation and the type of instruments for measurement—limnigraphs, flowmeter, turbidimeter, rain gauges, measurement stations in the river etc.; the strategy is to minimise the number of such instruments.

These instruments should, above all, be capable of recording data in groups and eventually for each site so that after a rainfall it is possible to analyse the impact and role played by each component of the network; this will require readjustment and revised instruction for the software of local automation (improved static management).

As the next step it is necessary to regroup the data at the central station and to establish a connection between this station and the works command so that modification of instructions becomes possible if dynamic management exists.

The cost of these instruments together with their installation and connecting them to a central station constitutes 1 to 2% of the total investment. It should also be kept in mind that preliminary modifications of storm weirs and the set of adjustable valves are imperative and that their cost (comprising study and modelling) is high—depending on the case, as much as 3 to 10 times the cost of the control devices.

For example, instrumentation for a small network with a central station, with 10 peripherals and 100 bits, requires an investment of 2 to 3 million francs; for a large network, say 300,000 inhabitants (1 PC, 50 peripherals and 1000 bits) the required investment would be 40 to 50 million francs: about 15 to 20% for probes, 20% for activators, 30% for transmission, 10% for organisation of the central station and 25% for changeover to dynamic management (local automation, decision-making personnel etc.).

Lastly the operational cost—electric power constitutes a small fraction but maintenance and breakdown costs are significant—must be included. Further, as already mentioned, surveillance equipment should be of good quality and sufficient in number.

If these investments are undertaken parallel with enhanced knowledge of the network and personnel experience, the result is economy of personnel and especially greater reliability and cost effectiveness. Further, changeover to improved static management and then to dynamic management assures a marked increase in the efficiency of storm works. This has been clearly demonstrated in simulations carried out on discharge of the catchment area of Bièvre in Ile de France. The changeover to dynamic management resulted in 10 to 20% reduction in volume of discharge into river Seine using the same works; reduction in pollution was also significant. By maintaining the same impacts on the natural environment, the capacity of the planned basin (530,000 m^3) could be reduced by more than 70,000 m^3, equivalent to an economy of more than 150 million francs, which in turn is several times the cost of equipment of regulation (at 1993 prices).

IAWPRC [72] describes state-of-the-art instruments for remote management and **Green** [73] discusses the advantages accruing therefrom, reinforcing the claims made above.

Valérie [71] has described the organisation of a system of this type at three levels, installed by the **District of Val de Marne**.

2.2 MODELLING OF NETWORK AND SIMULATIONS

The present discussion includes neither a description of the method of modelling currently in use (the reader may refer to Appendixes 8 and 9 in **Valiron**, ref. F) nor details of the major models available on hydraulics and transport of solids in conduits (the reader is referred to **Bertrand-Krajewski** [42]).

Most of the modern models use for the flow chosen, three different descriptions, listed here in the order of increasing complexity:

— **kinematic wave** in which only the friction and the gravity forces are considered; downstream effects are thus neglected;

— **advective wave:** in addition to frictional and gravity forces, a hydraulic gradient is also considered, thereby taking into account the limiting conditions downstream and simulating the real flow;

— **complete solution** of Barré de Saint Venant equations, simulating rapid variation of flow rate and level.

Many models include free surface and flows under pressure as well as overflow and storage. They take into account the singularities of the network and various structures for flow regulation, valves, weirs and pumps, and introduce the Q(h) laws for these works.

Some models include turbulent and streamline flows and their relationship. All models include a description of the water cycle: rainfall and loss by overland flow—employing both actual and fictitious rainfall.

Rain-water flow is described by various softwares enabling passage from hyetographs to hydrographs at the network outlet and at each node by selecting different forms of representation: curves, areas, time, simple reservoir or otherwise, double reservoir and in some cases the Caquot method which gives only the peak flow rate.

Figure 5.5 explains the passage from a hyetograph to various hydrographs on a network.

Some models, for example the American **SWMM**, comprise a block of computations enabling simulation of 9 pollutants and entrainment by rainfall of deposits at the bottom as well as the phenomena of deposition, resuspension and transport in the network. This model may be equipped with a block which computes the increment of pollutants in the storage works.

MOUSE, the Danish model, and its extension MOUSETRAP, is represented in France by **SAFEGE** [40] and computes the annual and maximum discharge into the natural environment (of BOD, SS, nitrogen, phosphorus) on the basis of a long series of historic rainfalls.

FLUPOL (ref. VI-6), an operational model perfected jointly by the Agence de l'Eau Seine-Normandie, the Compagnie Générale des Eaux and the Syndicat des Eau d'Ile de France, operates as a specific model closely integrating quality and quantity. It has been successfully used in the Parisian region and northern France. It is better suited for studying the transfer of suspended particles than those flowing along the bottom, the latter being better handled by SWMM. A succinct description is given by **Valiron** in Appendix 8 of ref. A.

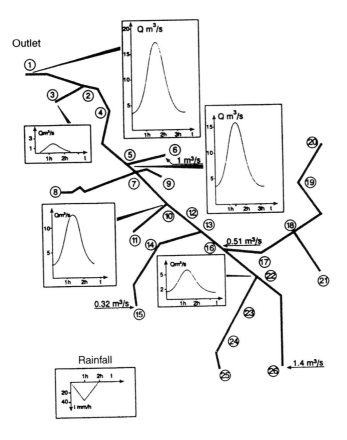

Fig. 5.5: *Transformation of hydrographs on a network.*

Some characteristics of the three most common models in France are given in Table 5.9.

RERAM	Does not take into account the influence of contaminants; the zone of contamination is not determined.	STU
CAREDAS	Zone of contamination determined but its influence upstream ignored.	SOGREAH
CEDRE	Zone of contamination determined and ,its influence upstream taken into account.	INSA Lyon

Table 5.9: *Some French models of hydraulic flow rate*

The model **CAREDAS** is used in particular at Nancy and Seine St. Denis and Paris, **MOUSE** in Hauts de Seine and Berne (Switzerland) and **FLUPOL** in Ile de France.

• **Calibration of models**

Without repeating the discussion of model calibration (which consists of adjusting it to several sets of data and comparing results with measured values), it may be recalled that the cost of such measurements and calibration are high, and the more complex the model, the longer the time required. It is desirable to choose a model compatible with the level of information required but it should be as simple as possible. As a matter of fact, complexity is not synonymous with accuracy.

For example, to study the storm pollution of the agglomerated network of Metz discharged in the Moselle and Seille, calibration of MOUSE, a relatively simple model, required measurement of discharge and pollution poured over 15 storm weirs during 11 storms and over 45 weirs during 2 storms as well as installation of 2 rain gauges; execution of this programme took 6 months—from April to September 1991 (**Safège** [43]).

• **Simulations**

Once the model has been calibrated, it can be used to determine what happens in the existing weirs during a particular storm or during several storms over a year. For example, **Degardin** [44] used Flupol to determine the volume of overflow during a uniform decennial storm on the Parisian agglomeration from storm basins (Fig. 5.6) as well as the quantity of SS.

Another use of a verified model is to compare, say, the volumetric discharge in a given weir and the flow rate and level along a sewer for a network in its normal state (Figs. 5.7 and 5.8) and, after providing a certain amount of storage upstream, a comparison of different storms.

From the volumetric inflow of water and pollution to a given weir, one can calculate the effective capacity and percentage of pollution reduction by using efficiency curves, such as those given in Fig. 4.19 of Chapter 4.

The SWMM model has a block for computation of these results by applying the EPA norms (see Annex. II).

Hypothesising extension of urbanisation upstream, one may also determine its effect due to an increase in capacity of transport, or discrete storage, or even discretisation of discharge on the plot.

The results of a model for different upstream discrete storage can be compared with those obtained with simulation of remediation of discharge for various storm basins.

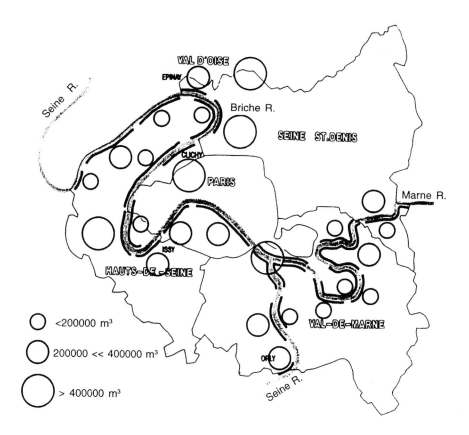

Fig. 5.6: *Simulation of overflow volumes from storm weirs of the Parisian agglomeration for a decennial storm.*

These simulations give a sufficiently accurate idea of the effects of inclusion in the network of a particular work and thus help in making an economic choice while assessing the cost of the relevant pre-projects.

To avoid carrying out a study of a large number of cases, effort should be made to eliminate some by utilising the knowledge acquired through analysis of construction of other networks.

2.3 SOME REMARKS ON CHOICE BY WAY OF CONCLUSION

• **Improvement of weirs,** which per se play a key role in the discharge, is given **due priority** in France as well as abroad, especially in England as noted by **Green** [19] and **Ellis** [11]. These authors emphasise that improvement should be effected in weirs with high setting level or lateral weirs with a settling chamber

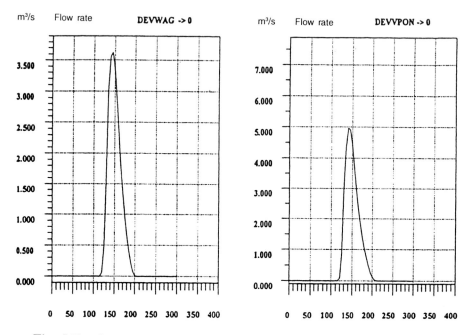

Fig. 5.7: *Curves of discharge in two adjacent weirs obtained from the model MOUSE for a given storm.*

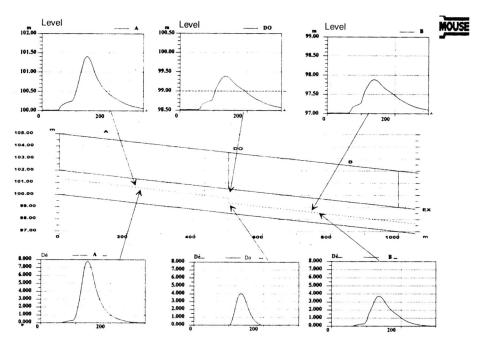

Fig. 5.8: *Curves of level and flow rate along a sewer for a given storm.*

108

or weirs with a vortex or centrifuge, giving preference to works not needing human intervention. It appears that **non-adjustable static works should be reserved** for smaller networks (some tens of thousands of inhabitants) or for works with small discharge and reduced impact on the environment of other networks.

With progress in automation and information, it appears difficult to avoid **equipment of such works with adjustable thresholds.**

As pointed out by **Hansen** [46], the wider employment of valves (or storm basins) which direct more pollution towards the treatment plant is related to expansion of its capacity. Gousailles (case study no. 9) indicates the broad lines of adaptation of instrumentation along the network at Achères for monitoring pollution during the rainy season.

• **To tackle problems arising from greater urbanisation,** it is today considered advisable that parallel to urban growth efforts should be made to restrict the quantity of rain-water flow to the original level. Necessary provision should be made in **regulations pertaining to sewerage in the community concerned** (see ref. g). Either lamination should be carried out on the site or the developer made to pay a suitable sum to the community to enable it to construct a storage-lamination basin at the downstream. Following the example of Bordeaux and Seine St. Denis, such provisions are currently underway throughout France.

• **Construction of storage-lamination basins** was originally conceived as a measure for limiting the diameter of the main sewers downstream. More recently in Germany (see **de Carné**, [36]), The Netherlands (see **Voorhoeve** [34]) and in England, such works have been planned to reduce the number of weir discharges and not expressly for the purpose of remediation, although they do serve this purpose to some extent. **It thus appears that the choice between storm basins directly concerned with pollution and storage-lamination basins concerned with frequency and quantity depends on the objective to be achieved.**

This means that if the storage capacity of the network, including the basins is small, the two solutions should be simulated and the better one implemented. **Breuil** (case study no. 9 and ref. 47) determined the better coherence between storage and decantation, which can be found by a judicious combination of the two types of basins by varying the rate of flow of the material deposited in two hours by lamination of the upstream basin, to the downstream decantation basin (Fig. 5.9).

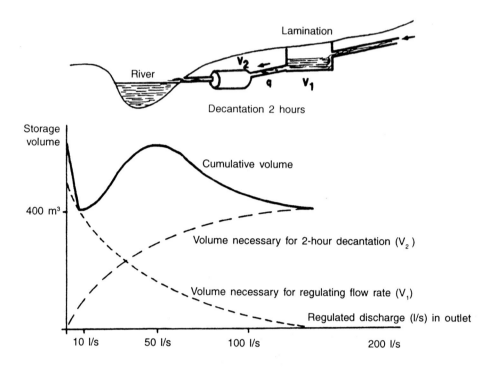

Fig. 5.9: Storage-decantation coherence.

Assuming that the cost per m³ of the two types is the same, one finds 10 l/s and 330 m³ for the storage and 70 for decantation—a ratio of 80 to 20%. These values are close to those found in the literature (the Dutch and the Germans speak of 2/3 and 1/3).

CHAPTER 6

Regulations Concerning Discharge of Weirs

1. SITUATION BEFORE DIRECTIVE OF CEE, NO. 91 271 OF 30 MAY 1991

Until 1960-1970 the entire world was unaware of the pollution carried by rainstorm run-off; it was considered clean water and hence the rather vague laws prescribed that the overflow from storm weirs be diluted with rain-water so as not to perturb the natural environment.

1.1 THE UNITED KINGDOM

The weirs were designed to discharge dry season flow (DSF) in the main sewer. The law permitted a dilution of 6; flow higher than 6 DSF was discharged and the rest transported to the treatment plant designed to treat 3 DSF. Water exceeding 3 DSF was stored in basins and diverted after the storm to the treatment plant equipped with an overfill weir.

Towards the 1960s, a relationship was formulated which took into account the water consumption of the population and of industry and the flow to be transported was given by:

$$D_t = DSF + 1350 \, P + 2 \, E$$

where DSF is the rate of flow during the dry season
   P the population
   E the mean intake in litres/day by industry
This is equivalent to the first formula corresponding to a daily consumption of 270 litres per inhabitant.

Later, the Water Authorities became aware of the pollution in rainstorm run-off and designed a new type of weir which could retain the pollution and prevent its discharge by means of a settling chamber or a high weir. This did not preclude overloading of weirs as a result of urban growth. For better control of pollution discharge, the weirs were subject to a declaration or authorisation.

111

The law concerning dilution (1 to 4 or 5) was prescribed in these countries also. Weirs situated in the catchment area were required to allow a flow rate of 15 l/s per impermeable hectare to the downstream (this flow rate was 8 to 9 times the peak flow rate of waste water for 50 m² impermeable surface per inhabitant). On the other hand, weirs situated more downstream on the main sewer leading to the treatment works cut down any flow rate higher than:

$2 Q_s + Q_f$ (Q_s the flow rate during the dry season and Q_f additional flow)

For further limitation of spread of pollution, the directive **ATV 117** [50] was promulgated which stipulated discretisation of peak flow by storage in the network. **In The Netherlands arrangements** were made for storage at different points of the network to reduce the frequency of overflow (first to 10 per year and later to 5).

However, awareness in the 1980s that the flow could be strongly contaminated led to the promulgation of directive **ATV 128** [37], which introduced a new concept: 'the networks should be so designed that 90% of the annual degradable matter will be treated at the biological section of the plant'.

Later, in 1994, the ATV directive was modified to treat the rainstorm run-off separately and take into account the norms of discharge of the plant.

The concentration of COD in the discharge from the treatment plant should be lower than 70 mg/l; hence dilution has to be calculated. For the combined overflow, the **annual average of COD** should be below 107 mg/l; taking into account the amount of storm water at the plant, the average is about 37 mg/l. A computation, rather complex, for an annual rainfall of 800 mm is given in **Annex. I (document 3).**

In The Netherlands, towards 1985, the computation of basins which corresponded to a storage of about 20 m³/impermeable hectare was replaced by the constraint of a surface load smaller than 5 m/h; the output of works so dimensioned corresponded to a reduction of 70% in annual flow.

1.3 FRANCE

Development has been similar but slower. Since 1949, computations have been in accordance with **circular CG 1333** (ref. d) based on Caquot's formula; the coefficients of concentration, slope, surface, rainfall being corrected by several factors such as basin shape, evaporation of rain-water, capacity of the network etc. But the objective of the circular was to limit the initiative of the designer to furnishing simple data in the form of charts regarding the flow rate Q to be taken into account and the sectional area of the main sewer.

Computations were based on experiments and measurements done on the main sewers in Paris. The weirs were operative even for rainfall beyond that taken into account.

In 1977, the above circular was replaced by a **new instruction 77.284** (ref: e) whose two major modifications are:

— introduction of detention tanks to reduce peak flow rate and thus the cost of the network; the earlier instruction considered only the basins upstream of the plant for limiting the peak capacity of treatment;

— replacement of computations based on the formula with revised coefficients by a calculation based on mathematical models for basins larger than 200 ha, especially the CERA model, now replaced by RERAM.

The reader desirous of more details on the genesis of these two instructions may consult **Dupuy** (ref. J) and **Fouquet** (ref. I) for an example of network calculations.

Lastly, the instruction leaves the flow rate at which the weir begins to overflow to the designer; the choice rests on economic considerations subject to the constraint that provision be made for dilution. Most of the French combined networks were designed for downstream flow rate of 2 to 3 DSF and the treatment device always provided with a weir for discharging flow in excess of its design capacity (very few devices were constructed for a flow rate higher than the DSF).

It may be noted that many networks carry parasitic water and drain much larger urbanised surfaces than those existing at the time of the project. This explains why many weirs operate daily during the peak dry season and the peak hot days.

2. NEW REGULATIONS

2.1 EUROPEAN DIRECTIVE OF 21 MAY 1991
CONCERNING TREATMENT OF RESIDUAL WATER

This directive (ref. b) is applicable to residual water such as household waste water or its mixture with industrial water or rain-water. It stipulates that such water should be treated after its collection in a tank. This obligation of treatment, whose degree is defined by a directive on treatment plants, pertains to all waste and storm water and must be implemented by 2000 or latest 2005.

The directive covers only such overflow on the route as can be regarded as corresponding to exceptional circumstances of a case of excessive cost. It authorises the member states to take necessary measures to limit the pollution resulting from a storm to a level lower than that in the DWF and allows only a few overflows per year. Thus it covers the provisions already made in some

113

countries of the European Community, in particular Germany, the Netherlands and Switzerland, and the provisions of the **overall management** recommended by the **new French law of January 1992** (ref. a) but gives them greater importance.

2.2 NEW REGULATORY PROVISOS IN FRANCE

• **Decrees 93 742 and 93 743** (ref. c) cover storm weirs under the rubric 520 pertaining to nomenclature; they are subject to authorisation or declaration depending on the daily discharge of pollutants as shown in Table 6.1.

By polluting flux is meant the flow carried by the main sewer over the weir. Thus the limits correspond to agglomerations of 600 to 3000 inhabitants. Whether authorisation or declaration is required depends on the quantity of discharge or ratio of flow rate to a reference flow rate as indicated in Table 6.2.

Daily polluting flux	
Larger or equal to 120 kg of BOD_5	Authorisation
Between 12 and 120 kg of BOD_5	Declaration

Table 6.1

Volume or rate of discharge	
$V > 10,000$ m³/day or $D_{day} > 25\% \ D_{ref}$	Authorisation
$2000 < V < 10,000$ m³/day or $5\% \ D_{ref} < D_{day} < 25\% \ D_{ref}$	Declaration

Table 6.2

By reference flow rate is meant the average montly DSF for a 5-year period recurrence (QMNA 5 years).

The volume of discharge is that corresponding to the capacity of the outlet.

If the dossier submitted for declaration is found to be complete and in order, a receipt is issued. The Comité Départemental d'Hygiène (CDH) may stipulate additional conditions. Authorisation is granted by the Prefect's office after a public inquiry and consultation with the CDH.

• **Decree 94 469** (ref. f) stipulates that after the inquiry, the community shall delimit:

    — zones of common sewerage in which the domestic waste water is to be collected;

— zones in which it is necessary to make special provision with respect to storm run-off to protect the aquatic environment;
— zones or individual sewerage systems whose monitoring is the responsibility of the community;
— zones in which soil impermeability should be monitored to reduce rainstorm contamination.

The Prefect delimits the parameters of the urban areas and defines by appropriate bylaw the goals of reduction of polluting substances which the community should fulfil through a sanitation plan.

• **Bylaw of 22 December 1994** (ref. i) specifies the principles of the prefectorial bylaw concerning reduction of pollution. In particular, the rainfall on which the characteristics of the network are based may change, thereby necessitating a dynamic approach. Requisite results are introduced into the collection system (average number of weirs, connections, rate of collection). Only the discharge larger than 100,000 needs to be continuously measured; for the others monitoring is done during the period of overflow and the quantum of pollution discharged estimated.

Articles 6 and 7 stipulate that the networks of collection—weirs and treatment plants of the same urban area—should be designed, constructed and managed in a consistent manner taking into account their effect on the receiving medium.

Exploitation should minimise the overall quantity of polluting material discharged in every operation.

For this purpose one may provisionally include in the treatment systems for **significant contamination, nominal amounts of contaminating substances** without endangering the safety of the system. Under these conditions of operation, the proprietor is not obliged to adhere to the stipulations of the bylaw of authorisation. Furthermore, other alternative or complementary solutions (detention or storage basins...) may be utilised.

• **As pointed out by Trionnaire [49], the 'Guide to recommendations for application of documents appended to the circular of 12 May 1995'** (ref. j) stipulates a progressive approach. The general approach (see **Deneuvy [82]**) consists of:

1) **Early construction of the works and installation of the equipment required for:**
— suppression of direct discharge of the dry season;
— regulation of weirs;
— specification of capacity of treatment plant;
— formulation of a policy of limitation of overland flow by controlling urbanisation or implementation of alternative techniques.

2) Simultaneously, **supervising the hydraulic operation of the sewerage system** so as to evaluate its effects in different meteorological situations of the annual cycle.

3) Determining at the end of this analysis or **as a second step, the storage facility to be installed** for taking into account the rainfalls of low frequency of return (generally, of the order of monthly precipitation); the second step should be planned outside the schedule chalked out for the directive.

4) **Formulating, after analysing the results obtained and carrying out necessary observations**—an ambitious strategy—the priority of construction determined in terms of the per unit cost of eliminating pollution corresponding to each facility or work (coss-effective ratio vis-à-vis receiving environment).

An overall approach at the level of a catchment area of the receiving environment is necessary to ensure the coherence of the actions undertaken in this plan, in accordance with the existing documents on planning (SDAGE, SAGE, Directive Schemes of Sewerage, Plans for Urbanisation etc.).

In general, emphasis should be given to immediate effects, especially the impact of discharge of organic contaminants, the source of anoxia in the receiving environment. To this end the locations of discharge should be judiciously chosen to preclude any shock effects. Nevertheless, other effects should not be ignored in an overall strategy, in particular the impact of heavy metals, organic micropollutants and nutrients in zones sensitive to eutrophication.

• **The provisos for application of these texts** thus emphasise the diagnostic study prescribed in article 16 of the decree under reference regarding the objectives of pollution control of the prefectorial bylaw cited above. However, this study carried out at the urban area level, has to be cross-referenced with the action taken with respect to the catchment area.

This approach motivates the community to study the incidence of storm events for at least one year, considering in particular the immediate effects.

Various approaches are possible depending on the size of the agglomeration:

— **for small- or medium-size urban areas** the concept of controlled degration of the quality of the natural environment may be used, allowing violation of the objectives of quality 10% of the time, by accepting declassification of a class;

— **for larger urban areas,** the basic consideration should be maintenance of a sufficient level of dissolved oxygen for preservation of fish life, which requires a rather complex study.

These two approaches are not mutually exclusive. Section 3 of this chapter contains a more complete description of the policy to be adopted with respect to objectives of quality.

The enhanced costs inherent in such a scheme mandate a **step-by-step implementation** keeping financial considerations in mind. POS have to be revised to incorporate provisos for monitoring overland flow and their costs duly calculated if satisfactory progress of the planned programme is to be realised.

2.3 New british policy

Implementation of the Directive has been detailed in a document prepared by the 'National Rivers Authority' in consultation with the government and the water industry. This document considers the following:
— rate of dilution,
— storage capacity of the network,
— limitation of the number of discharges.

Due attention is given to the receiving environment and thus to the objectives of quality based on water usage.

Three distinct criteria have been formulated according to water use:
— aquatic life in inland watercourses,
— coastal waters for swimming,
— aesthetics of coastal and inland waters.

Based on these usages, norms regarding the discharge and quality of the environment to be maintained were formulated (**NRA** [48]).

• **Norms pertaining to quality to be maintained in inland waters** for various periods of exposure are given in simplified form in Table 6.3.

Frequency	Concentrations of non-ionised ammonium (NH_4-N mg/l)		
	1 h	*6 h*	*24 h*
1 month	0.150	0.075	0.030
3 months	0.225	0.125	0.050
1 year	0.250	0.150	0.065
Frequency	Concentrations of dissolved oxygen (DO mg/l)		
	1 h	*6 h*	*24 h*
1 month	4.0	5.0	5.5
3 months	3.5	4.5	5.0
1 year	3.0	4.0	4.5

Table 6.3: Norms for quality of inland water for various periods of exposure

• **Norms for water for swimming**

— Concentration of faecal coliforms higher than 2000/100 ml allowable during less less than 1.8% of the swimming season (or 2 days out of 120 days).
— No more than 3 storm discharges during the 120-day season.

• **Norms for aesthetics** are given in Table 6.4

Category of aesthetic aspect	Frequency of discharge	Norm of expected discharge
High	> 1 per year ≤ 1 per year	Separation of solids of 6 mm Separation of solids of 10 mm
Medium	> 30 per year ≤ 30 per year	Separation of solids of 6 mm Separation of solids of 10 mm
Low and ex-category		Good engineering design

Table 6.4: Norms for discharge for protection of aesthetics

NRA imposes design criteria for weirs for the separation of solids according to the level of aesthetics desired. According to the norms listed in Table 6.3, the procedure to be adopted differs with the amount of discharge (Table 6.5).

Small discharge	Exercise control of discharge Dilution > 8:1 No interaction with other discharges
Medium discharge	Use simple techniques of evaluation of impact Dilution < 8:1 Limited interaction with other discharges Population > 2000 Carp
Large discharge	GPU procedure with accurate modelling Dilution < 2.1 Interaction with other discharges Population > 10,000 in terms of equivalent inhabitants Carp or salmon

Table 6.5: Accepted methods for determining impacts in inland waters according to amount of discharge

It may be noted that the approach is similar to that recommended in France; it requires studies on models (see Sec. 3) for large discharges.

The techniques to be used are detailed in Table 6.6 (according to the amount of discharge).

Small discharge	Methods requiring minimum data (e.g., formula A)
	Coastal and estuarine waters not including zones for swimming and/or zones of aquaculture
Medium discharge	Hydraulic methods of collection with evaluation of frequency of discharge
	Population 2000-10,000
	Affects zones for swimming and/or zones of aquaculture
Large discharge	Complex methods (hydraulic model of main sewer with evaluation of frequency of discharge and option of model of coastal dispersion and evaluation of impact)
	Population 2000-10,000 in terms of equivalent inhabitants
	Affects zones for swimming and/or zones of aquaculture

Table 6.6: Accepted methods for discharge from weirs in inland and coastal waters according to amount of discharge

The GPU procedure (see **Rumsey** [51]) is briefly described in the following section.

3. POLICY REGARDING OBJECTIVES OF QUALITY AND RAINSTORM DISCHARGE

It is desirable to fix the level of quality in the river to be maintained in case of storm discharge, focusing logically, as in the case of waste water, on concentration of dissolved oxygen resulting from the effects of numerous parameters. As mentioned in Sec. 4.1 of Chapter 1, the smaller the duration of exposure, the better the tolerance of fish; therefore storm duration must be taken into account.

Another important factor is the period of storm return for which proper protection must be provided because the cost of controlling the amount of pollutants increases rapidly with length of period of return.

Figure 6.1 explains the Danish recommendations for the lowest level of dissolved oxygen concentration to be maintained during rainstorm discharge. This required level decreases with the period of storm return. As pointed out by **Harremoes** [55], the curves were drawn based on these considerations.

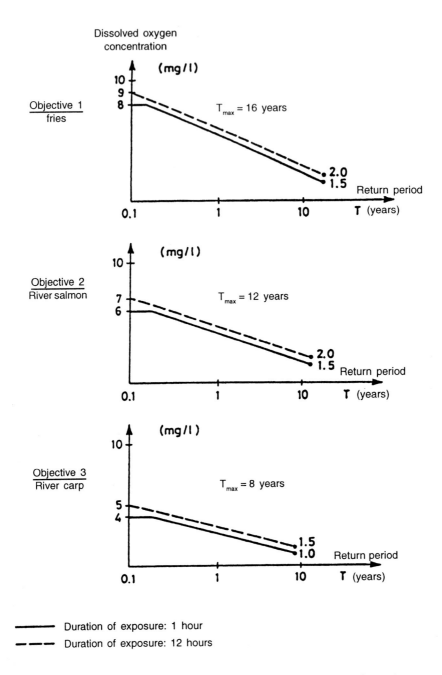

Fig. 6.1: *Danish criteria for objectives of quality during the rainy season.*

The objective of river quality has been defined: a computational procedure to determine the allowable discharge needs to be developed and the means for reducing actual discharge ascertained.

Aside from determining the quantum of BOD_5 in the discharge, attention must be paid to its impact on the dissolved oxygen in the river. The methods for studying this impact are those used in the case of discharge of waste water. **But it must be borne in mind that the impact of a short-duration discharge, though not harmful to fauna—as illustrated in Table 1.13 of Chapter 1— must be taken into account while carrying out computations.** This topic and the methods developed in England are discussed in Section 3.2 below.

3.1 METHOD OF CONTROLLED DEGRADATION OF WATER QUALITY

As discussed above, this is a matter of so determining the arrangements that a rainfall of rank 9 (10% of 90 days) downgrades the quality of the watercourse to the permissible lower limit, while a rainfall of rank 10 does not (or hardly) affect the receiving medium. A second condition is that a rainfall of rank 5 should not entrain a downgrading to more than two classes.

Safège applied these rules to the agglomeration (urban area) of Mulhouse and obtained the figures given in Table 6.7.

	Basin volume		Calculated volume
Catchment area	Rainfall of rank 5	Rainfall of rank 10	By ATV method
B 01	450 m³	200 m³	108 m³
B 04	180 m³	48 m³	38 m³
B 07	170 m³	70 m³	36 m³

Table 6.7

One can readily see that the rainfall of rank 5 (similar to that of 3/4 months return period and 2 hours duration) is more restrictive than the criterion of rainfall of rank 10.

Lastly, it can be seen that this method of downgrading is stricter than the corresponding ATV method, especially with respect to the totally stored monthly rainfall.

3.2 COMPUTATIONAL METHODS FOR REDUCTION OF
DISSOLVED OXYGEN

The method employed in the United Kingdom is based on the Danish work mentioned in Sec. 3. The Water Research Centre (WRC) of the Urban Pollution Management, UK constructed the chart given in Fig. 6.2.

For a storm of annual frequency, the concentration of dissolved oxygen should be maintained at 2.5, 3.5 and 4.0 mg/l for durations of 3, 6 and 24 hours respectively. The instantaneous minimum of 2 mg/l is well above the 0.5 mg/l requisite for the survival of rainbow trout.

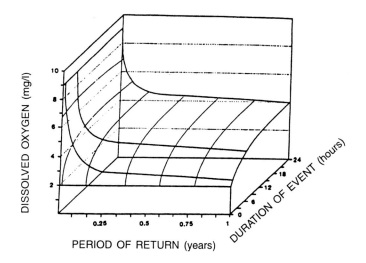

Fig. 6.2: Three-dimensional criteria for dissolved oxygen (in mg/l) for an intermittent discharge as a function of the return period and duration of discharge.

In the same vein of thought, the WRC tried to perfect the models of impact by combining its MOSQUITO model with the Danish method of impact, MIKE II. The model STOAT, which simulates treatment in the plant enables better simulations.

3.3 COMPARISON BETWEEN METHOD OF DOWNGRADING
AND COMPUTATION OF DISSOLVED OXYGEN

Studies carried out at Metz (see Sec. 3, Chapter 1) by **Safège** [43] showed that the **two methods were quite similar** because the rainfalls considered were similar.

122

As a matter of fact, the method of downgrading based on rainfalls of rank 9 was conducted on basins storing monthly rainfall to conserve Seille, effluent of the most sensitive Moselle, while in the software MIKE II the most downgrading tracer NH_4 was applied to the volumes of the basin very close to those of the other method.

3.4 MODELS OF RIVER QUALITY

Most of these models are one-dimensional and many assume a steady flow and pollution transported without dispersion. The model of the Seine at Montereau in Paris, developed on this basis at the Agence de l'Eau Seine Normandie, uses the equations of the oxygen cycle; the parameters modelled are dissolved oxygen, BOD, ammonium and nitrates.

Based on his successful experience in Kalito, the reactions were modelled as shown in Fig. 6.3.

Other models simulate the phenomena in transitory flow. This is the case of **MIKE** II (see **Safège** [40]), developed by the Danish Institute of Hydraulics and used in France, consisting of separate modules for simulating flow in rivers from the overland flow in the basin and changes in pollutants. It reproduces the conditions of low water and flooding. Supplementary modules include the transport-dispersion, transport of sediments and simulation of water quality.

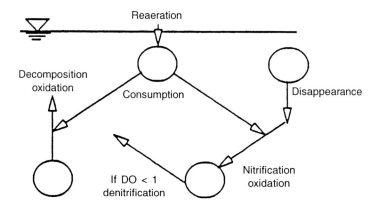

is particularly applicable to large rivers during the period of small flow when sensitivity to perturbation of discharge is highest. It simulates the biodegradation and reaeration using the same laws as used in the Kalito model, in particular the **Gameson** formula, and thus uses the same parameters for verification of these phenomena.

For the study of impact, the discharge of some weirs of the drainage network of the Parisian agglomeration is currently used (see **Degardin** [44]).

MOKA-EAU, developed by **Hydratec** [78], is currently used by SIAAP on the rivers Seine and Meurthe (see Sec. 3.6).

3.5 MODALITIES OF APPLICATION

The modalities developed in the UK as a follow-up of the European Directives are described here; they have been in use for the past two years and represent an advancement over those developed in France.

Defect/ Means of control		Network	Overflow	Pollution of receiving medium
Means of inspection		Defective inspection of sewer	WASP-SIM	
Matrix of progressive examination		Length of defective sewer	Zone of overflow	Perturbing storm discharge
Action to be taken *on the sewer* —reduction of flow				
rate	M	x	xx	xx
—repairs	f	xxx	--	--
—renovation	M	xxx	x	x
on the weirs	f	--	xx	xxx
—reduction of	x	xx	xx	
discharge	M			
—maintenance	f	--	x	x
—reinforcement	F	--	xxx	xx
—replacement	F	xxx	xxx	xxx

Table 6.8

Table 6.8, taken from **Ellis** [30] who cites **Clegg** [57], shows that a very pragmatic and progressive approach has recently been attempted for rehabilitation of networks and weirs. This approach is currently being pursued after the success of the research programme initiated in 1986 for the **Management of Urban**

124

Pollution (see **Rumsey** [51]); this programme has enabled use of the following tools of modelling of the components of collector systems which discharge in the natural environment:

stormal: entry of storm events in hydraulic models

mosquito: simulation of the quality of water in sewers

stoat: simulation of treatment plants

mike II: assessment of impact on the watercourse

quests: coastal dispersion of bacteria

These tools have been tested by actually applying them to study modᵃlities of operations (**Crabtree** [53]) described in the Urban Pollution Management Manual. This manual incorporates the progress in weir design (see **Balmforth** [54]) and suggests a unique approach to planning, using the tools in the order of their complexity and cost and covering the following phases:

Phase A: pre-project

Phase B: collection of data and tools

Phase C: development of solutions

Phase D: selecting the appropriate solution

This approach may involve hard work in the collection of data and break-up of cost of works but this cost of the study is more than compensated by the resultant economy.

In this context, **Crabtree** [31] discusses the case of development of a storm weir discharging at the downstream, water from a town of 15,000 inhabitants. This discharge created an inadmissible degree of pollution and therefore a project of storage was undertaken. The volume calculated from only the hydraulic analysis of the sewer was 1800 m3. Use of the procedures given in the Urban Pollution Management Manual, involving the impact of the discharge, enabled restricting the storage to 1000 m^3, resulting in 40% reduction in the total cost as summarised in Table 6.9.

The reader interested in details of the British methods of control of storm discharge is referred to Urban Pollution Management (ref. L), which contains several examples of computation and employment of various models mentioned above.

3.6 EXAMPLE OF APPLICATION IN FRANCE—CASE OF NANCY

The method briefly described in Sec. 3.1 was applied to the Meurthe River. The model Moka-Eau was used in a series of measurements taken on the river and the discharge of the urban area of Nancy. Downgrading of the Meurthe due to rainy episodes was 60 days in summer from class 1 B to 2 and 10 days from class 1 B to 3; thus fish mortality was high during the low flow rate and high temperatures. The objective of the developments attempted was to abide by the

Design	Storage (in m³)	Cost of study in million francs	Cost of construction in million francs	Total cost in million francs
Traditional	1800	0.5	7.5	8
New approach	1000	1	4	5
Saving of 40% in total cost				

Table 6.9

constraints of a downgrading limited to 10% of the duration for a rank one or 5% for rank two, such that there would be no fish mortality for a frequency of more than once in ten years.

The study compared the three solutions for the discharge, namely, storage, treatment of the water stream and storage, and lamellar decantation, and found the last was the best.

According to **Hydratec** [78], 8 basins would need to be constructed to fulfil the objective, with a total capacity of 50,000 m³ and treatment at the rate of 10 m³/s. Working of a basin is schematised in Fig. 6.4.

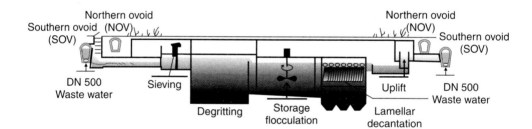

Fig. 6.4: Working of proposed basins.

A pilot basin of 7000 m3 capacity and 3 m³/s rate of treatment will be installed first. The associated catchment area will cover 660 ha (of which 260 ha will be impermeable) and the population 40,000.

A sketch of this basin together with the sites at which measurements are to be taken is shown in Fig. 6.5. It highlights the enormous effort required for studying about ten possible modus operandi by providing two storage compartments with eventual flocculation and two cross-current or counter-current lamellar decanters as well as several sites where measurements are to be done. The estimated cost of this work is 40 million francs.

126

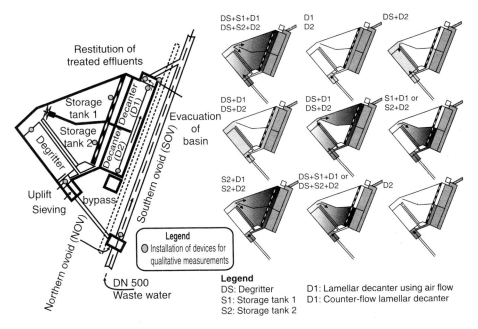

Restitution of
treated effluents

Storage
tank 1

Storage
tank 2

Decanter (D1)

Decanter (D2)

Evacuation
of
basin

Degritter

Uplift
Sieving

bypass

Southern ovoid (SOV)

Northern ovoid (NOV)

DN 500
Waste water

DS+S1+D1
DS+S2+D2

D1
D2

DS+D2

DS+D1
DS+D2

DS+D1
DS+D2

S1+D1 or
S2+D2

S2+D1
S2+D2

DS+S1+D1 or
DS+S2+D2

D2

Legend
◯ Installation of devices for
qualitative measurements

Legend
DS: Degritter
S1: Storage tank 1
S2: Storage tank 2

D1: Lamellar decanter using air flow
D1: Counter-flow lamellar decanter

Fig. 6.5: Installation and possible modus operandi.

3.7 CURATIVE ACTIONS IN NATURAL ENVIRONMENT

3.7.1 Reinforcement of natural self-purification

In view of the high cost of works for reduction of inflow of pollution during a storm, efforts have been made worldwide to improve the natural process of self-purification of the natural environment by artificially supplementing the oxygen supply. The set-up in London together with brief references to the set-up at Ruhr (Germany) and the one planned for Paris are described below.

Since 1980 researchers in London have been studying the effect of injection of oxygen in the Thames to offset the effect of deoxygenation caused by inflow of rainstorm run-off. Tests undertaken from a barge capable of injecting 10 tons of oxygen per day led to the **construction of a special boat 'The Thames Bubbler'** 50 m long, 10 m wide, displacing 592 tons, which has been in service since 1989. This boat has a plant which produces oxygen from air and utilises the selective properties of some absorbants. The plant comprises three units of compression-decompression separating nitrogen and oxygen by means of 'Xeolite' and supplies 30 tons of oygen of 93% concentration per day. This oxygen is injected in the water by a system shown in Fig. 6.6 by means of injectors producing fine bubbles and a cross-flow of 2 m³/s of highly oxygenated pressurised water.

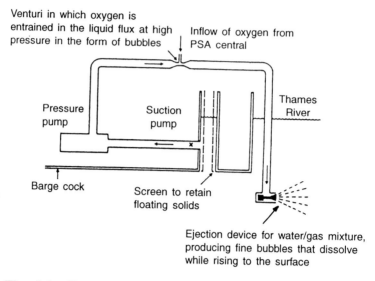

Venturi in which oxygen is entrained in the liquid flux at high pressure in the form of bubbles

Inflow of oxygen from PSA central

Pressure pump

Suction pump

Thames River

Barge cock

Screen to retain floating solids

Ejection device for water/gas mixture, producing fine bubbles that dissolve while rising to the surface

Fig. 6.6: *Sketch of equipment of the barge 'Fine Bubbles'.*

The boat is soundproof and belongs to 'Thames Utilities'. It is equipped with 3 diesel engines of 1050 kilowatts for oxygen production and is propelled by two 'Aquamaster' pumps with orientable jets so that it has good manoeuvrability. It operates under the command of the 'National River Authority' and is placed at the lower point of the oxygen sag curve (5 to 6 times a year) as determined by 8 probes along the river.

Construction of a second boat is presently under consideration, which will have fixed stations along the banks and use hydrogen peroxide.

At Ruhr in Germany, installations for reoxygenation operate at hydroelectric plants; injection is done through turbines and a boat is also used on some lakes.

In Ile de France, since 1970-75 the Department of Navigation has been carrying out tests from rafts equipped with air diffusers. SIAAP is currently collaborating with **Air Liquide** to study means of injection of pure oxygen.

3.7.2 Reduction of river flotsam

Floating barrages can be used to collect flotsam; the collected material is then despatched to a centre where it is destroyed. This procedure is implemented by SIAAP on the Seine in close collaboration with the Department of Navigation and the Communes of barrage installations. They block only a fraction of the river so as not to disturb navigation and are stationed at places where wastes accumulate naturally[1]. Collection is done by a special boat 'Port Marly', 20 m

---

[1] Colour plate no. 14.

long, 5 m wide, equipped with a rolling belt with metallic mesh trailing in the water and a hydraulic arm for manipulating gross waste. The operation was started in 1988 with two barrages. Today seven barrages are in service and 12 projected by 1995.

The cost of floating barrages is of the order of 200,000 francs without value added tax and that of flotsam removal and transport (not including destruction) is 2000 F/ton.

SIAAP [77] extracted 755 tons of wastes in 1994 using 6 barrages. When the strength of the fleet grows to 12, it is expected that the amount of waste removed will be 1500 tons/year since a study revealed that the total waste flowing in the river will be of the order of 6000 tons/year. It was further estimated that only one-third of this quantity will be contributed by sewers.

CHAPTER 7

Conclusions and Suggestions for Strategic Action and Enhancement of Knowledge

One of the objectives of this monograph is to make the reader aware of the importance of storm weirs (and direct pollution due to rainstorms) in the design and management of a sewerage network.

The 1992 law regarding water and the recent decrees on weirs stipulate the overall management of water in urban areas and emphasise the need for recourse to SAGE in the case of larger zones.

Various solutions presently available and described in this book should be adapted to each of the numerous cases encountered; their success depends, of course, on an in-depth understanding of the working of the network and its impacts on the natural environment during the rainy season.

In this last chapter a strategy is proposed for activating each network and remedying any malfunctioning. Lastly, steps to be taken in future to reduce the lacunae in our knowledge are suggested.

1. STRATEGY FOR ACTION

1.1 Inventory of weirs and bypasses

Such an inventory should not be limited to just weirs and the discharge from pumping stations operated in the case of flooding, but should include the works of exchange among the main sewers (bypasses, pumping stations etc.) because these works modify the catchment area of weirs and thus influence their performance; they also help to increase the storage capacity of the network by using meshworks.

• This inventory should be completed in the form of a catalogue with a file for each weir or bypass, indicating its location on the network plan.

Each file must include a sketch showing the following details:

— characteristics of the upstream and downstream sewers (diameter or sectional area, slope etc.);

- characteristics of the outlet with reference to the natural environment (diameter, slope, length);
- description of the natural environment (level during low water, in flood etc.);
- type of weir (high-low, frontal, lateral etc.) and the threshold (length etc.);
- auxiliary equipment (screen, harrow etc.), mechanism of regulation, settling chamber, storm basin etc.;
- instruments for measurement.

These data enable computation (or evaluation) of the maximum flow rate during a rainy event, taking into account $[Q_{1p} + Q_{2p}]$ and the amount transported downstream (Q_{3p}) and hence the amount of overflow, $Q_{max\ p}$ (see Fig. 7.1).

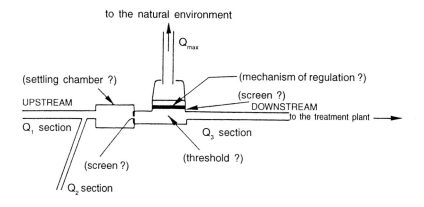

Fig. 7.1: Sketch of components of a weir.

Determination of flow rate can be a rather tricky affair when a weir is located downstream of a zone in which sewerage is partly separated or when bypasses can modify the catchment area depending on their location. These problems are more likely to appear, however, if the catalogue has been prepared for operation from upstream to downstream. This corresponds to the situation described in Fig. 3.14 of Chapter 3 [or the case of Bièvre (see **Valiron,** case study no. 5)].

• Lastly, the file should include the data available on the discharge (functioning during the dry season, steady, occasional, number of discharges in a year etc.) as well as calibrations and any complaints made by the inhabitants.

Aided by this inventory, the weirs and bypasses should be classified according to the river or sewer in terms of the maximum discharge during the rainfall under consideration. The catchment area of each weir should concomitantly be specified (the same catchment area may serve several weirs).

Breuil (case study no. 9) discusses as an example the catchment area Vieille Mer-Moreé in Seine Saint Denis to demonstrate the complexity of configuration of networks consisting of both separate and combined systems and numerous bypasses. He also points out that intermediate weirs play an important role in modifying inflow and that the inventory must not be limited to just the downstream weirs in the Seine.

Even the concept of dry and rainy seasons is modified by a sewer that can transport, at a given moment, the effluents of the dry season because it plays the role of a meshwork and hence is the carrier of rainstorm run-off.

The foregoing discussion emphasises the need for complete analysis of the network, including all the points of discharge.

When no catalogue exists, or the one that does is incomplete (lacks detailed information), steps should be taken to fill in the missing data. An example of such a catalogue prepared in 1995 by the Val de Marne is shown in Fig. 7.2. Completion of such a file may involve extensive topographic work if the original plans have been lost. Nevertheless, **knowledge of the hydraulic components of this inventory** (or description of sewers and location and size of the overflow weir) as well as the extent of the catchment area, is **indispensable for modelling the network. This highlights the importance of combining the two approaches.**

1.2 Imbrication of weirs in the network

In the preceding chapters it was conclusively shown that the efficiency of weirs is related not only to actions on the weirs per se, but also to the creation of storage capacity upstream of the network or on the plot of land itself to reduce the inflow of pollution, or further to the provision of means of treatment of discharge by the storm basins or stream of water. It was further emphasised that the role of storage in the network per se enhances with the meshing of main sewers by means of bypasses directing excessive rainstorm run-off in one sewer towards another.

Lastly, the pumping stations located near the lowest weirs, which cannot function during floods, prevent the local overflow of water carried by the network during the rainy season.

Maintenance of the network itself prevents sedimentation (or limits it) and thus helps in conserving its transport capacity. In the absence of proper maintenance, the water stream rises without change in flow rate and hence the weirs operate more often with higher volume and discharged pollution.

Table 7.1, taken from AGHTM [64], illustrates this imbrication which leads to a search for improvement of weirs in the framework of the network.

1.3 Co-ordination of weirs and treatment plants

As demonstrated by researchers such as **Hansen** [46], the improvement of storm

INSEE 42	Commune: Joinville-Le-Pont		N° EA09 42DO.01	N° DEX 121

Location: Quai du Barrage-Rue Beaubourg	Sec 4	Type DO

Nature	Section	Work
Departure UN	1.80/1.00	42401 (UN)
Arrival Rain-water	1.80/1.00	42401 (Rain-water)

Final destination: Marne River	Code discharge: 05.01.36

Conditions of operation:	Barrage with girders and valve—isolating valve on 42401 (Rain-water) before discharge in Marne River
Special equipment before final destination	C.R.P upstream and removal of floating matter of dry season towards Work XII

Date: 1991	Scale: 1:500

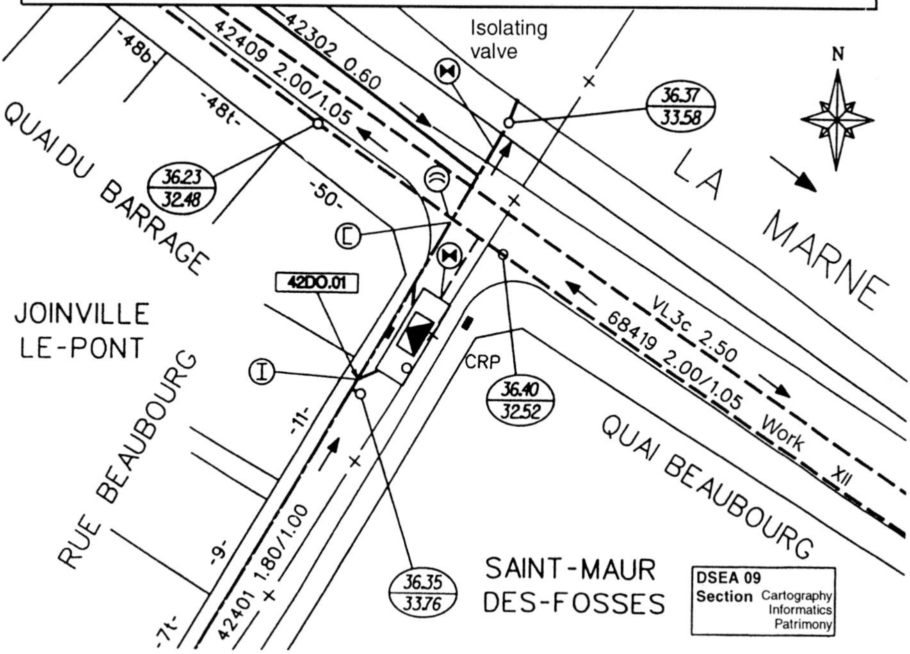

Fig. 7.2: A sample from the catalogue of Val de Marne.

134

weirs, in particular through storage, should be closely co-ordinated with the eventual increase in the capacity of the treatment plant which has to handle a larger amount of pollution; this has also been pointed out by **AGHTM** [64].

Management of works of discrete detention or storage-decantation is a major problem in the exploitation of sewerage systems during the rainy season. In fact, the capacity of the plant leads to a reduction of efficiency when the flow rate is higher than a prescribed limit. Also, the evacuation of these works should not affect the functioning of the treatment plant and reduce its capacity, thereby creating an unwarranted collection of pollution, as shown in Fig. 7.3. In the example under consideration, while reducing the quantum of pollutants discharged at the level of weirs by increasing the storage volume, the quantum of pollutants discharged at the treatment device increases due to increase in the volume treated and degradation of the pollutants. The overall quantum of the discharged pollutants thus reaches an asymptote; it may even increase beyond this limit.

As already discussed in Sec. 2.3 of Chapter 5, **Gousailles** (case study no. 8) has detailed the possibilities of treatment at Achères during the rainy season by combining a physicochemical treatment and using tertiary treatment with a biofilter during the dry season; execution is planned for 2001. This method will

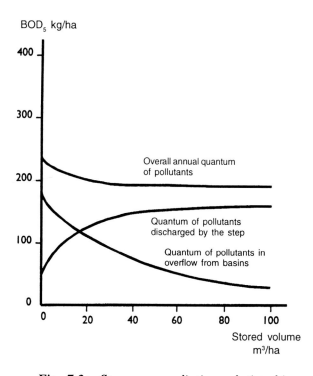

Fig. 7.3: Storage-remediation relationship.

Works	Actions	Goals	Constraints	Complementary actions
	Elimination of discharge during dry season.			
	Optimisation of calibration of overflow layer	Reduction of volume of overflow	Hydraulic limit attained too soon	Storage indispensable
Storm weirs	Reduction of number of weirs and equipment in adjustable overflow weirs.	Centralisation of discharge in view of treatment or storage		Centralised control
	Elimination of macrowastes by screens or baffles	Reduction in discharge of wastes	Maintenance	
Storage Laminate storage	For good efficiency, indispensable for a weir, lamination basin, storm basin	Reduction in volume discharged	Maintenance	
Meshing the network	Bypass between sewers	Reduction in number of discharges and volume of overflow	Risk of overflow	Centralised control
Pumping station in case of flood	During flood part of water from lower portion of network to be pumped towards the river	Battle against overflow		
Treatment of water stream	–storm basin (storage-decantation) –laminar decantation	Reduction of quantum of pollutants discharged	Concentration of effluents by regrouping DO	Storage plug
Reduction of pollution in overland flow	Disconnect rain-storm run-off at level of plot of land	Reduction in volume of overflow		
Maintenance of network	Preventive treatment of network Improvement of hydraulics during dry season. Installation of chamber for resuspension of deposited matter.	Reduction in resuspension of deposited matter. Prevention of untimely discharge		Informatics maintenance

Table 7.1: Modus operandi of network

be thoroughly tested as part of a study which takes into account the impact of rainstorm basins, whose design depends on whether the retained pollution is transported to the Seine River after discrete decantation or transferred after the storm over Achères.

1.4 DISCUSSION OF WEIRS AND DIAGNOSTIC STUDIES OF NETWORKS

Table 7.2, taken from **Lyonnaise des Eaux** (ref. G), shows the four phases of a diagnostic study of an existing network, revealing the malfunctions and suggesting remedial measures; it covers the numerous problems related to weirs and their improvement.

The first step is to consider the data available on rainfall and catchment area, the weirs, rentention basins and storm basins as well as the works located on the plot. Subsequently, attention should focus on treatment during the rainy season and the impact of storms on the natural environment.

With respect to measurements, those made during the rainy season are difficult to carry out and require higher expenditure. The same is true for the programme concerning control during the rainy season and determination of the type of works to be installed (retention basins, storage basins, lamination of inflow on the land, weir equipment) and priority with respect to their installation.

Modelling the network is indispensable for determining the types of works and their effect on the network per se as well as on the natural environment. This may be considered the preliminary step towards development of a permanent model of the network's operation and its effect on the external environment.

On the other hand, some parts of the diagnosis, such as search for parasitic water and the study of waste water and operation of the network and treatment plant during the dry season, are not directly related to the weirs and their improvement, but their cost is hardly a fourth of that of the complete diagnosis.

Thus this analysis shows that whenever financially feasible, the means of improvement should be considered both for the rainy and dry seasons. Attention should focus in particular at this stage on all aspects of the weirs (especially an accurate description and precise dimensions).

It may also be noted that modelling of the network can be undertaken and executed by using the approximate relationships for $Q(h)$ pending accurate measurements; the model suffices for estimating the storage in the network and calibration and regroupment of weirs. It is also possible to simulate the effect of replacement of the weirs based on the approximate relationships for $Q(h)$ by a modernised weir (or weirs) obeying an accurate $Q(h)$ law.

Obviously, the cost of diagnostic studies varies strongly with the data collected to date (in particular, the topography of the network) and the accuracy desired in determination of the works to be undertaken. According to the statistics of

Collection and use of data	Data relating to collection	Waste water { industrial / population / consumption of potable water → theoretical waste water pollution / → theoretical flow rate of waste water. Rainstorm run-off { rainfall characteristics of basins / impermeable surface area
	Data relating to equipment of sites and transport	Data on state and functioning ↓ Dossier of review and recommendations { −works on upstream land / −main sewers / −pumping station / −weirs / −degritters, grease removers / −retention basin / −storm basin
	Data relating to treatment	Balance of treatment { dry season / rainy season. State of functioning of treatment works
	Data relating to receiving environment	Existing quality { dry season / rainy season. Desired quality { dry season / rainy season
Measurements rainfall-pollution flow rate	Dry season	on waste water sewer and combined sewer (high water table) → { permanent parasitic / inflow. on rainstorm run-off sewer and overflow from combined sewer → { direct discharge of / waste water
	Rainy season	on waste water sewer → bad connection. on combined sewers and eventually rainstorm run-off sewers → { direct discharge of / waste water
	Nocturnal inspections	to identify sections producing permanent parasitic water
Specific studies	Televised inspections	on sections producing permanent parasitic water
	Smoke tests	for identifying run-off connections on waste water
Hierarchical programme of rehabilitation	Hierarchy in terms of....	−m³ eliminated parasitic water / −m³ waste water restored in network / −pipeline of run-off water disconnected from waste-water network / −reduction in flux of pollutants discharged in receiving environment } expressed in equivalent population

Table 7.2: Four phases of diagnostic study

the Agence de l'Eau Seine-Normandie, the average cost per inhabitant varies from 150 F for small towns to about 20/25 F for sizable cities (Fig. 7.4) with a margin of ± 25% (1993 prices).

These diagnostic studies also lead to development of sewerage schemes for municipal or metropolitan areas, as in the case of the **Seine Saint Denis** district [69].

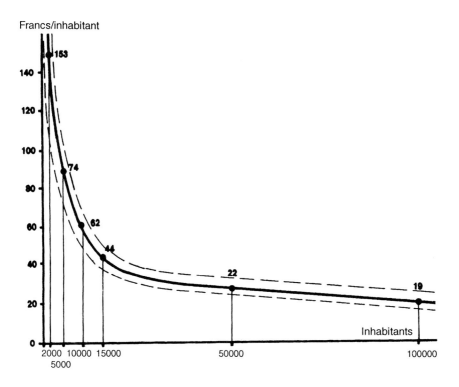

Fig. 7.4: Cost of diagnostic studies.

1.5 TWO EXAMPLES OF OVERALL STRATEGY— ARRAS AND SEINE ST. DENIS

By carrying out measurements on the network and one of the weirs during the dry and rainy seasons using the programme Flupol, the **District of Arras** [79] showed that for storm events of trimestrial return, the discharge represented 80% of the pollution of the dry season of the agglomeration. Given this observation, a programme was launched to determine the most economical solution for reduction of the pollution discharged, by combining various possibilities: extension of the capacity of the treatment works during the dry and rainy seasons, decantation

139

basin evacuating towards the natural environment or towards the network, extension of sewerage system to zones in which it did not exist, and improvement in the transfer of waste water. The least expensive process of removing 1 kg BOD_5 or TKN involves the use of a storage basin evacuating in the network, followed by one discharging in the natural environment and lastly treatment of the rainstorm run-off (in the ratio 1 : 1.5 : 3).

• **Case of Seine Saint Denis** (Breuil, case study no. 9)

In case study no. 9, **Breuil** elaborates the strategy of 'Seine Saint Denis' for reducing the impact of discharge in the natural environment, which differs in the dry and rainy seasons.

> • During the dry season discharge from the works is banned during the period May to October—the critical period for the Seine River; all available means are used to divert the flow, including that from storm works, towards Achères.
> • During the rainy season different strategies are used for different catchment areas, with storage capacity as the criterion in choosing the appropriate strategy:
>> – **For Moreé Sausset,** the capacity of basins is adequate for reducing the discharge and directing the maximum possible volume to Achères.
>> – **For the combined zone towards the Seine,** it is not possible to direct 80 m³/s of the decennial rainfall towards Achères because the capacity of the sewer is 12 m³/s. The problem was solved by intercepting the overflow on the banks and treating it in plants of large capacity. The first step in this direction was installation of a basin of 180,000 m³ capacity under the parking lot of the Grand Stadium.
>> – **For discharge in the Marne River,** where the extant storage capacity is small, it is planned to construct basins to reduce inundation and to increase the capacity at the neighbouring treatment works of Noisy le Grand so that it can handle 30% of the decennial flow discharge.

1.6 STORM WEIRS AND SAGE

As shown earlier, notably in Sec. 3 of Chapter 6, improvement of weirs for an urban centre and reduction of the polluting flux should be closely co-ordinated with those of similar works of other communities which use the same natural environment as the outlet.

These ideas regarding the goals and the steps to be undertaken to achieve them can now be planned under the direction of SAGE, set up by law in 1992 to deal with problems pertaining to development and management of water.

2. IMPROVEMENT OF KNOWLEDGE

2.1 BETTER KNOWLEDGE OF EACH NETWORK

It is not proposed to repeat here discussion of the diagnostic study presented in the preceding section; attention is focused instead on analysis of the past performance of works. Knowledge of the malfunctions and defects in such works is useful for calibrating models to tackle problems relating to such topics as overflow and fish mortality.

The knowledge gained from such analyses and the information gathered from the models already in use and the data collected, are of immense value for developing, step by step, an ideal model and for accurate planning of future works. Experience shows that such analysis is also helpful with reference to the financial aspects of the works to be undertaken.

2.2 GENERAL STUDY

Various lacunae in the knowledge gained from the discussions included in this book can be slowly filled in by study and research at home and taking due note of the progress reported in other countries. Some important points in this regard are listed below.

• **Better assessment of the cost of different works,** paying special attention to high costs of foundations; also consideration of environmental constraints, especially with respect to:
— valves of weirs
— storm basins,
— discrete decanters
— lamination works on the upstream land
— screens and skimmer.
Of equal importance to better cost assessment is estimation of the expected output of each of these works vis-à-vis reduction of polluting flow.

• **Perfection of methods of assessment of impact of discharge** in the natural environment and downstream, as well as improvement of the models of computation of the works of discharge.

• **Assessment of cost of study and measurements** to be undertaken for improvement of works and equipment as well as performance of the network:
— calibration of weirs,
— installation of measurement devices,
— introduction of a centralised control of management.

• **Economical and technical study** of the best means for reducing river flotsam (trap, screen at the weir, floating barrage). The study should be detailed with respect to various available techniques, their drawbacks and merits.

• **Comparison** of various methods of computation of the capacity of basins of decantation (EPA method, ATV method etc.) and traditions in force in France. The comparison should involve three or four cases and only the measured efficiency considered.

ANNEXURE I

DOCUMENTS CONCERNING
THE METHODS OF COMPUTATION

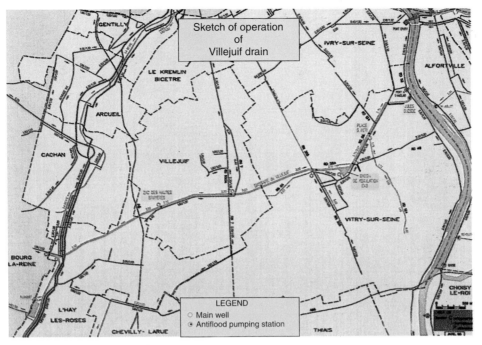

1 Sketch of operation of Vitry network (case study no.4)

2 Vitry control basin (case study no.4)

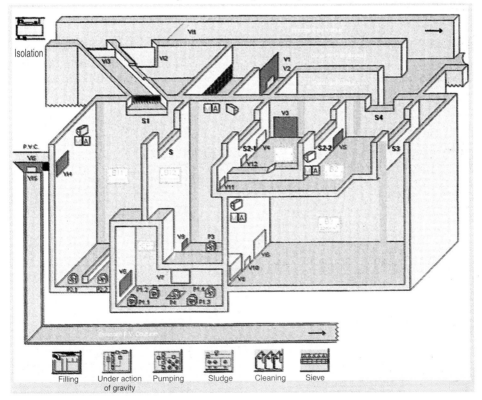

Filling | Under action of gravity | Pumping | Sludge | Cleaning | Sieve

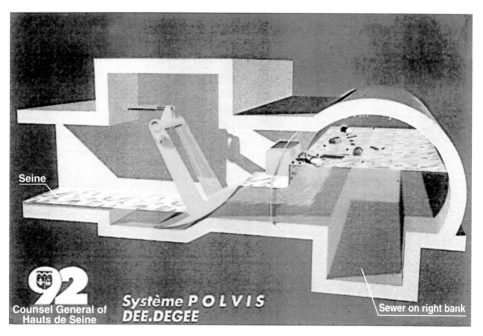

3 Polvis system with overflow and retention of wastes (case study no.6)

4 Construction of control stations on the Seine River, ADES network

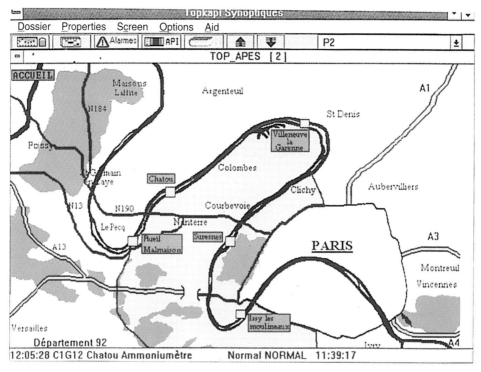

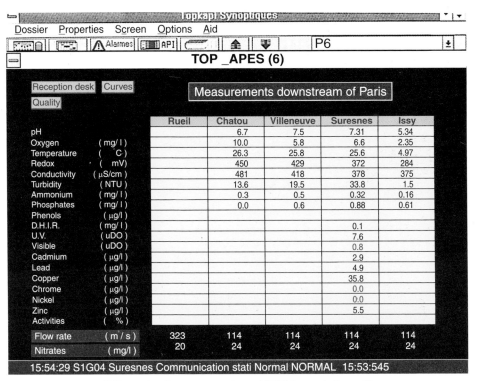

Dossier   Properties   Screen   Options   Aid

[icons] Alarmes API P6

TOP _APES (6)

Reception desk | Curves
Quality

Measurements downstream of Paris

		Rueil	Chatou	Villeneuve	Suresnes	Issy
pH			6.7	7.5	7.31	5.34
Oxygen	(mg/ l)		10.0	5.8	6.6	2.35
Temperature	(C)		26.3	25.8	25.6	4.97
Redox	(mV)		450	429	372	284
Conductivity	(μS/cm)		481	418	378	375
Turbidity	(NTU)		13.6	19.5	33.8	1.5
Ammonium	(mg/ l)		0.3	0.5	0.32	0.16
Phosphates	(mg/ l)		0.0	0.6	0.88	0.61
Phenols	(μg/l)					
D.H.I.R.	(mg/ l)				0.1	
U.V.	(uDO)				7.6	
Visible	(uDO)				0.8	
Cadmium	(μg/l)				2.9	
Lead	(μg/l)				4.9	
Copper	(μg/l)				35.8	
Chrome	(μg/l)				0.0	
Nickel	(μg/l)				0.0	
Zinc	(μg/l)				5.5	
Activities	(%)					
Flow rate	(m / s)	323	114	114	114	114
Nitrates	(mg/l)	20	24	24	24	24

15:54:29 S1G04 Suresnes Communication stati Normal NORMAL 15:53:545

5 Example of measurements on ADES (case study no.6)

6 Control station of ADES network (case study no.6)

7 Flood control station of l'Avre at Boulogne (case study no.6)

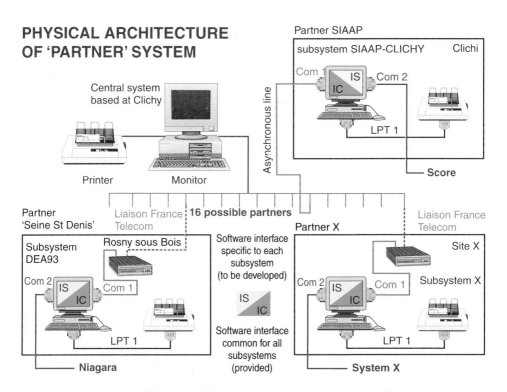

PHYSICAL ARCHITECTURE OF 'PARTNER' SYSTEM

Partner SIAAP

subsystem SIAAP-CLICHY   Clichi

Com 1   IS   Com 2
         IC

LPT 1

Score

Central system based at Clichy

Printer   Monitor

Asynchronous line

Partner 'Seine St Denis'

Liaison France Telecom   **16 possible partners**   Liaison France Telecom

Subsystem DEA93

Rosny sous Bois

Software interface specific to each subsystem (to be developed)

Partner X

Site X

Com 2   IS   Com 1
         IC

IS
     IC

Subsystem X

Com 2   IS   Com 1
         IC

Software interface common for all subsystems (provided)

LPT 1

LPT 1

Niagara

System X

8. Architecture of 'Partner' system (case study no. 7)

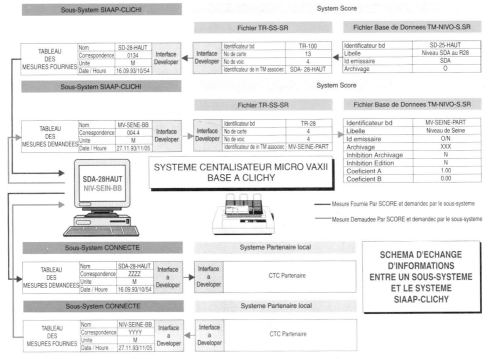

9 Sketch of exchange of information between a subsystem and the SIAAP-CLICHY (case study no.7)

10 Automatic cleaning of underground basin of Sevran (95) using a hanging bucket (photo: Counsel General 1989)

11 Cleaning of underground basin by tracto-shovel (CUB Bordeaux)

12 Permeable parking lot (document Counsel General 95)

13 Structure of overdimensioned sewers
(document Counsel General 93)

14 Floating barrage collector of wastes at Sevres (photo SIAAP)

15 Installation of measurement devices in a sewer (photo Safege)

16 Overflow in sewer towards an outlet (Paris : photo Safege)

Tables 1 and 2 Giving Values of the Coefficient m in the Bazin and Rehbock Formulas

Water depth (h) in (metres)	Height of sill (p) in metres								
	0.20	0.30	0.40	0.50	0.60	0.80	1.00	1.50	2.00
0.10	0.462	0.450	0.445	0.442	0.440	0.438	0.437	0.436	0.436
0.12	0.463	0.449	0.443	0.439	0.437	0.434	0.433	0.431	0.431
0.14	0.466	0.450	0.442	0.438	0.435	0.432	0.430	0.428	0.427
0.16	0.470	0.452	0.443	0.437	0.434	0.430	0.428	0.426	0.425
0.18	0.474	0.454	0.444	0.438	0.434	0.429	0.427	0.424	0.423
0.20	0.478	0.457	0.446	0.439	0.434	0.429	0.426	0.423	0.422
0.22	0.482	0.460	0.448	0.440	0.435	0.429	0.426	0.422	0.421
0.24	0.486	0.463	0.450	0.442	0.436	0.430	0.426	0.422	0.420
0.26	0.490	0.466	0.452	0.443	0.437	0.430	0.426	0.422	0.420
0.28	0.494	0.489	0.454	0.445	0.439	0.431	0.427	0.421	0.419
0.30	0.497	0.472	0.457	0.447	0.440	0.432	0.427	0.421	0.419
0.32		0.475	0.459	0.449	0.442	0.433	0.428	0.421	0.419
0.34		0.478	0.462	0.451	0.444	0.434	0.428	0.422	0.419
0.36		0.481	0.464	0.453	0.445	0.435	0.429	0.422	0.419
0.38		0.484	0.467	0.455	0.447	0.436	0.430	0.422	0.419
0.40		0.487	0.469	0.457	0.449	0.438	0.431	0.423	0.419
0.45		0.493	0.475	0.462	0.453	0.441	0.433	0.424	0.419
0.50			0.481	0.468	0.458	0.444	0.436	0.425	0.420
0.55			0.486	0.472	0.462	0.448	0.439	0.427	0.421
0.60			0.491	0.477	0.466	0.451	0.442	0.428	0.422

Table 1: Bazin's coefficient m as a function of h_0 (water depth) and h_s or p (height of sill)

Water depth (h) in (metres)	Height of sill (p) in metres									
	0.10	0.20	0.30	0.40	0.50	0.60	0.80	1.00	2.00	3.00
0.02	0.451	0.446	0.444	0.443	0.443	0.442	0.442	0.441	0.441	0.441
0.04	0.442	0.431	0.428	0.426	0.425	0.424	0.423	0.423	0.421	0.421
0.06	0.446	0.430	0.425	0.422	0.421	0.420	0.418	0.418	0.416	0.416
0.08	0.454	0.433	0.426	0.422	0.420	0.419	0.417	0.416	0.414	0.413
0.10	0.463	0.437	0.428	0.423	0.421	0.419	0.417	0.415	0.413	0.412
0.12	0.473	0.441	0.430	0.425	0.422	0.419	0.417	0.415	0.412	0.411
0.14	0.483	0.445	0.433	0.427	0.423	0.420	0.417	0.415	0.412	0.410
0.16	0.493	0.450	0.436	0.429	0.424	0.422	0.418	0.416	0.412	0.410
0.18	0.503	0.455	0.439	0.431	0.426	0.423	0.419	0.417	0.412	0.410
0.20	0.513	0.460	0.442	0.433	0.428	0.424	0.420	0.417	0.412	0.410
0.22		0.465	0.445	0.436	0.430	0.426	0.421	0.418	0.412	0.410
0.24		0.470	0.449	0.438	0.432	0.427	0.422	0.419	0.412	0.410
0.26		0.475	0.452	0.440	0.434	0.429	0.423	0.420	0.413	0.410
0.28		0.480	0.455	0.443	0.435	0.431	0.424	0.421	0.413	0.411
0.30		0.485	0.459	0.445	0.437	0.432	0.425	0.421	0.413	0.411
0.32			0.462	0.448	0.439	0.434	0.427	0.422	0.414	0.411
0.34			0.466	0.451	0.441	0.435	0.428	0.423	0.414	0.411
0.36			0.469	0.453	0.444	0.437	0.429	0.424	0.415	0.412
0.38			0.473	0.456	0.446	0.439	0.430	0.425	0.415	0.412
0.40			0.476	0.458	0.448	0.440	0.432	0.426	0.416	0.412
0.45				0.465	0.453	0.445	0.435	0.429	0.417	0.413
0.50				0.471	0.458	0.449	0.438	0.431	0.418	0.413
0.55					0.463	0.453	0.441	0.434	0.419	0.414
0.60					0.468	0.458	0.444	0.436	0.420	0.415
0.65						0.462	0.448	0.439	0.422	0.416
0.70						0.466	0.451	0.442	0.423	0.417
0.75							0.454	0.444	0.424	0.418
0.80							0.457	0.447	0.425	0.418

Table 2: Rehbock's coefficient m as a function of h_0 or H_e (water depth) and H_s or p (height of sill)

The values are valid for rectangular weirs with thin walls and no lateral contraction.

Tables 3 and 4 Showing Flow Rate in Litres / Second Per Unit Length (Metre) of Threshold Length as Given Respectively by the Bazin and Rehbock Formulas

Water depth (h) in (metres)	Height of sill (p) in metres								
	0.20	0.30	0.40	0.50	0.60	0.80	1.00	1.50	2.00
0.10	64.7	63.0	62.3	61.9	61.6	61.3	61.2	61.1	61.0
0.12	85.3	82.7	81.5	80.8	80.4	79.9	79.7	79.4	79.3
0.14	106.2	104.4	102.6	101.5	100.9	100.1	99.8	99.3	99.2
0.16	133.2	128.1	125.5	124.0	123.0	122.0	121.4	120.7	120.5
0.18	160.2	153.7	150.2	148.1	146.8	145.3	144.5	143.5	143.2
0.20	189.3	181.0	176.6	173.9	172.1	170.0	168.9	167.7	167.1
0.22	220.2	210.2	204.6	201.2	198.9	196.2	194.8	193.1	192.4
0.24	253.0	241.0	234.2	230.0	227.2	223.8	221.9	219.7	218.8
0.26	287.6	273.6	265.5	260.3	256.9	252.7	250.3	247.5	246.4
0.28	323.9	307.8	298.2	292.1	288.0	282.9	280.0	276.5	275.1
0.30	361.8	343.6	332.5	325.4	320.5	314.4	310.9	306.6	304.9
0.32		380.9	368.3	360.1	354.3	347.2	343.0	337.9	335.7
0.34		419.8	405.6	396.1	389.5	381.2	376.2	370.2	367.2
0.36		460.1	444.2	433.5	426.0	416.4	410.7	403.6	400.5
0.38		502.0	484.3	472.3	463.8	452.8	446.3	438.0	434.4
0.40		545.2	525.8	512.4	502.9	490.5	483.0	473.5	469.3
0.45		659.4	635.3	618.3	606.0	589.6	579.6	566.5	560.6
0.50			752.9	732.1	716.7	696.0	682.9	665.7	657.8
0.55			878.2	853.4	834.8	809.2	792.9	770.9	760.5
0.60			1 011.1	982.1	960.0	929.2	909.3	881.9	868.7

Table 3

Water depth (h) in (metres)	Height of sill (p) in metres									
	0.10	0.20	0.30	0.40	0.50	0.60	0.80	1.00	2.00	3.00
0.02	0.435	0.431	0.430	0.429	0.429	0.429	0.429	0.429	0.429	0.429
0.04	0.437	0.426	0.423	0.422	0.421	0.421	0.420	0.420	0.420	0.420
0.06	0.446	0.428	0.422	0.420	0.419	0.418	0.418	0.417	0.417	0.417
0.08	0.456	0.432	0.424	0.421	0.418	0.418	0.417	0.416	0.415	0.415
0.10		0.437	0.427	0.422	0.420	0.418	0.417	0.416	0.415	0.414
0.12		0.442	0.430	0.424	0.421	0.419	0.417	0.416	0.414	0.414
0.14		0.448	0.434	0.427	0.423	0.420	0.417	0.416	0.414	0.413
0.16		0.453	0.437	0.429	0.425	0.422	0.418	0.416	0.414	0.413
0.18		0.459	0.441	0.432	0.427	0.423	0.419	0.417	0.414	0.413
0.20			0.445	0.435	0.429	0.425	0.420	0.418	0.414	0.413
0.22			0.449	0.438	0.431	0.427	0.421	0.419	0.414	0.413
0.24			0.452	0.441	0.433	0.429	0.423	0.419	0.414	0.413
0.26			0.456	0.444	0.436	0.430	0.424	0.420	0.414	0.413
0.28			0.459	0.446	0.438	0.432	0.425	0.431	0.415	0.413
0.30				0.449	0.440	0.434	0.427	0.422	0.415	0.413
0.32				0.452	0.443	0.436	0.428	0.423	0.415	0.413
0.34				0.455	0.445	0.438	0.429	0.424	0.416	0.413
0.36				0.457	0.447	0.440	0.431	0.426	0.416	0.413
0.38				0.460	0.449	0.442	0.432	0.427	0.416	0.414
0.40					0.452	0.444	0.434	0.428	0.417	0.414
0.45					0.457	0.449	0.438	0.431	0.418	0.414
0.50						0.453	0.441	0.434	0.419	0.415
0.55						0.458	0.445	0.437	0.420	0.416
0.60							0.448	0.440	0.422	0.416
0.65							0.452	0.442	0.423	0.417
0.70							0.455	0.445	0.424	0.418
0.75							0.459	0.448	0.426	0.419
0.80								0.451	0.427	0.420

Table 4

Computation of Decantation Conditions in a Storm Basin

1. INTRODUCTION

Two methods of computation of conditions of decantation in a storm basin are given below, as presented in 'Stormwater Detention for Drainage, Water Quality and CSO Management' by **P. Stahre and Ben Urbonas** (ref. K):
- method of surface load,
- method recommended by the Environmental Protection Agency in 1986.

These two methods relate exclusively to the trapping of fine sediments.

2. SURFACE LOAD THEORY

2.1 BASIC RELATIONS

According to this classical theory, the depth of the basin has no effect on the process of sedimentation. But this theory is applicable only when the flow is uniform, steady and streamline—conditions not fulfilled in storm basins.

Doblin (1944) and Camp (1946) obtained an analytical expression for the sedimentation in a turbulent flow.

Figure 1, taken from their work, shows the relationship among three non-dimensional parameters:

$$\frac{V_s}{Q/A}, \frac{V_s \cdot H}{3\varepsilon} \text{ and } E \tag{1}$$

where

V_s is the velocity of sedimentation of a particle;

Q/A   load/unit surface area;

H   water depth in the basin;

ε coefficient of diffusion;

E   efficiency of sedimentation (fraction of trapped sediments).

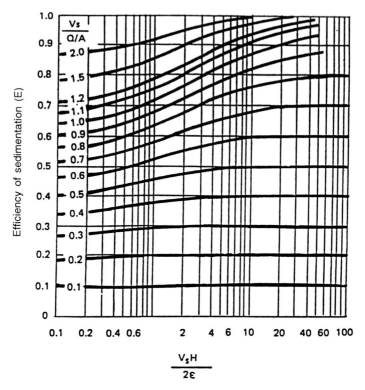

Fig. 1: *Diagram for estimating sedimentation.*

With the help of Fig. 1, the parameter E for particles of various sizes can be estimated. It may be noted that the sedimentation is influenced by the flow conditions in the basin.

Small values of $V_s \cdot H/2\varepsilon$ correspond to a turbulent flow and large values to a streamline one.

When $V_s \cdot H/2\varepsilon$ increases, E increases to approach the value

$$E = \frac{V_s}{Q/A} \qquad (2)$$

which is given by the classical theory of Hazen.

If the velocity distribution in the basin is parabolic, the coefficient of diffusion may be expressed as (Camp, 1946):

$$\varepsilon = 0.075 \sqrt{\frac{\tau}{\rho}} \qquad (3)$$

where τ is the critical shear stress, and
        ρ the density of water.

148

The critical shear stress depends on the size and density of the smallest particle susceptible to being mobilised by flow in the basin.

In Figure 2 the values of τ for silicon particles are shown as a function of the particle size and the purity of water (US Bureau of Reclamation, 1973). It may be noted that τ does not decrease when the particle size becomes very small (effect of forces of cohesion among very fine particles). Utilisation of τ values at the extreme left of Fig. 1 would be a rather conservative approach.

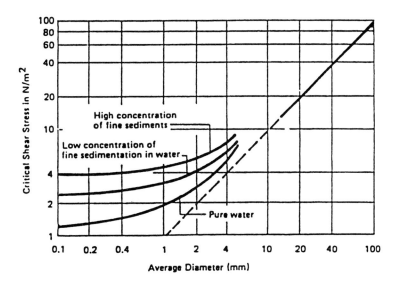

Fig. 2: *Critcial shear stress of entrainment of granules (diameter d) at the bottom of a canal.*

If the velocity of sedimentation V_s is known for particles of every size contained in the sediment load of a storm crossing the basin, E can be estimated for each class from Fig. 1. This can also be done for every value of Q/A.

Figure 3 presents the results of such computations under the hypothesis that $V_s \cdot H/2\varepsilon = 0.1$ (turbulent flow) and the assumption that only gravity and drag forces come into play during the process of sedimentation (for particles < 5 to 10 µm this hypothesis becomes controversial since electrostatic forces also play a role).

2.2 SUGGESTED PROCEDURE

(i) Determine the size and volume distribution of particles contained in the polluted storm-water.

(ii) Establish for each size the proportion of particles whose decantation is necessary for attaining a given water quality.

149

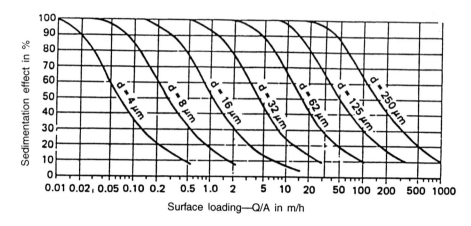

Fig. 3: Relationship E-Q/A for particles of different sizes.

(iii) Make a preliminary estimate of the maximum value of Q/A possible from Fig. 3.

(iv) Use Fig. 2 to estimate the velocity of sedimentation for the representative particles of every size.

(v) Using Fig. 1 and the data obtained from steps (i) to (iv), compute the value of E for each representative particle size.

(vi) Compute the overall efficiency by expressing the results obtained in (v) in terms of the size distribution of particles.

(vii) If the overall efficiency is unsatisfactory, repeat computations starting with a smaller value of Q/A in step (iii).

(viii) When a satisfactory value of the overall efficiency is obtained, determine the surface area and configuration of the basin.

The above procedure is easily programmed. The degree of accord between the results obtained and an actual situation has yet to be examined, however. Be that as it may, better results admittedly could be obtained were the following conditions fulfilled:

— satisfactory geometrical configuration,
— adequate dissipation of energy at the basin inlet,
— good knowledge of grain size-contaminant relationship,
— calibration of the model.

3. DIMENSIONING OF BASINS IN ACCORDANCE WITH EPA RECOMMENDATIONS OF 1986

The procedure is an adaptation of the probabalistic methodology originally formulated by Di Toro (1979).

150

3.1 RELATIONS APPLICABLE TO BASINS OPERATING UNDER DYNAMIC CONDITIONS

The approach used for estimating the conditions of sedimentation in these conditions is the same as that described in Sec. 2. It is based on the Fair and Geyer equation (1954):

$$R_d = 1.0 - \left[1.0 + \frac{1}{n} \frac{V_s}{Q/A} \right]^{-n} \tag{4}$$

where:

R_d = is the solid fraction decanted when the basin operates under dynamic conditions,
V_s = velocity of sedimentation of particles,
Q = peak flow rate,
A = surface area of the basin,
n = index of turbulence used for specifying performance of the basin.

Fair and Geyer (1954) suggest the following values of n:
n = 1: mediocre performance,
n = 3: good performance,
n > 5: very good performance,
n = ∞: ideal performance.

When n → ∞, eqn (4) reduces to the familiar expression (decreasing exponentially):

$$R_d = 1.0 - e^{-kt} \tag{5}$$

where

$k = V_s/h$, is the coefficient of sedimentation rate,
h = is the mean depth in the basin,
$t = V/Q$ is the detention time
V = is the volume of the basin.

The solution of eqns (4) and (5) can be represented in a graph similar to that given in Fig. 1.

By carrying out computation for a range of values of V_s and Q/A which are representative of the standard conditions of overland flow, one obtains the curves given in Figs. 4a and 4b.

Figures 4a and 4b respectively show the exact solutions of eqn (4) for n = 3 (i.e., good conditions of sedimentation) and n = 1 (bad or mediocre conditions).

The second solution may be more realistic in the actual conditions (variable incoming flow rate and depth).

Decantation under dynamic operating conditions of the basin depends on the intensity of rainfall. Hence estimates made in terms of seasonal or annual

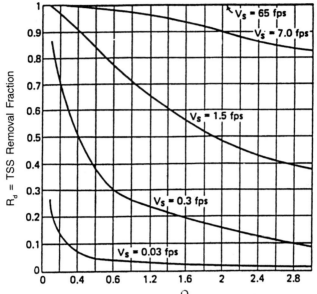

Fig. 4a: $\dfrac{Q}{A}$ *versus* R_d *for n=3*

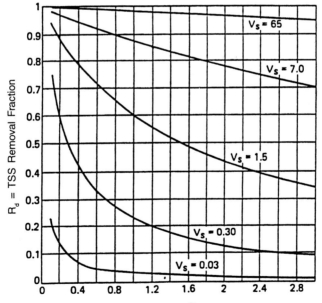

Fig. 4b: $\dfrac{Q}{A}$ *versus* R_d *for n=1*

averages have to be corrected to obtain an accurate assessment of the real fraction of SS decanted over a long period of time.

The long-term rate of removal may be estimated by using eqn (6); its graphic solution is given in Fig. 5:

$$R_L = Z \left[\frac{r}{r - \ln \dfrac{R_M}{Z}} \right]^{(r+1)} \tag{6}$$

where:

R_L: is the fraction decanted over a long period,

R_M: the average fraction decanted during the storm episode,

$r = 1/CV_Q^2$,

CV_Q = the coefficient of variation of the volume of flow,

Z = maximum fraction decanted.

In practice, the value of Z may be assumed to lie between 80% and 100% for all classes of particle size.

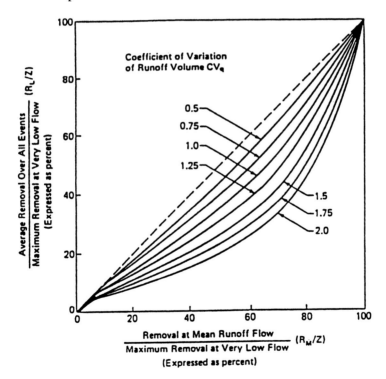

Fig. 5: *Long-term performance in dynamic conditions when decantation is sensitive to flow rate.*

153

However, when the finest sediments are electrostatically charged (clay), the value of Z lying between 10 and 50% is more appropriate for particles whose velocity of fall lies between 0.3 and 0.03 ft/h.

3.2 RELATIONS APPLICABLE TO QUIESCENT CONDITIONS IN THE BASIN

Sedimentation in the basin continues after the storm episodes subside and quiescent conditions are restored. The conditions of decantation then depend on the ratio of volume of basin to that of the preceding flow.

The rate of decantation is approximately given by the following equation:

$$R_Q = V_S A_B \qquad (7)$$

where:

R_Q = is the fraction decanted in the quiescent condition,

V_S = the decantation velocity of a particle in ft/h,

A_B = the surface area of the basin in ft^2.

The characteristics (date of starting, duration) of quiescent periods constitute random parameters. This fact is taken into account in Fig. 6 for making a practical estimate of R_Q.

First of all the ratio of the effective basin volume and the mean run-off volume (V_E / V_R) should be determined; the effective basin volume is not equal to the total volume of the basin and depends on the characteristics and the chronology of storms.

The ratio of the effective basin volume and the mean run-off volume can be estimated from Fig. 7. To make use of this Figure, the mean volume of run-off received by the basin has, however, to be known and this volume can be estimated by means of a model of the rainfall-run-off relation using the official meteorological data.

Thereafter the ratio E of the volume of SS decanted in quiescent conditions to the run-off volume is estimated as:

$$E = \frac{U_M R_Q}{V_R} \qquad (8)$$

where

U_M is the mean interval between storms (h$_2$),

V_R the mean run-off volume (ft^3) over the period of study.

From E and the ratio of basin volume to mean run-off volume the required ratio—effective basin volume/mean run-off volume—can be obtained by using Fig. 7 and the overall rate of decantation computed thereafter by means of Fig. 6.

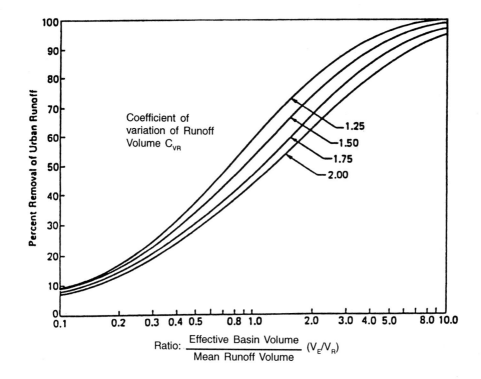

Fig. 6: *Mean long-term performance of a retention basin under quiescent conditions.*

3.3 RELATIONS APPLICABLE TO COMBINED CONDITIONS

A procedure for estimating the total fraction of SS decanted during a succession of rainy and quiescent periods is given below.

Let f_D be the fraction of time during which the basin operates in dynamic conditions

$$f_D = \frac{D}{U_M} \tag{9}$$

The corresponding fraction of time for quiescent conditions can therefore be expressed as:

$$f_Q = 1.0 - f_D \tag{10}$$

where

D is the mean duration of rainy episodes,

U_M the mean interval between two successive episodes.

155

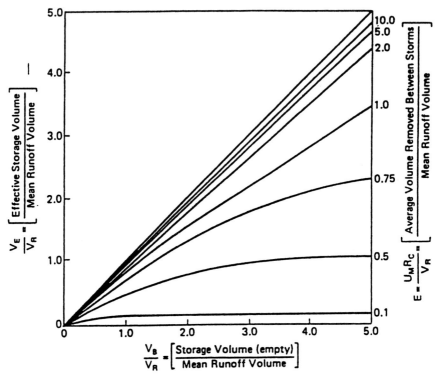

Fig. 7: *Effect of precedent storms on long-term performance in quiescent conditions (EPA 1986).*

The SS entering the basin is first subjected to a decantation in dynamic conditions.

The process of sedimentation in quiescent conditions thereafter takes place on a fraction of the volume remaining in the basin after the end of the storm and is related to the part of SS which is not decanted in dynamic conditions.

The total decanted fraction is thus given by the relation

$$f_T = 1.0 - F_D F_Q \qquad (11)$$

where

F_D is the fraction not decanted in dynamic conditions,
F_Q the fraction not decanted in quiescent conditions.

German Method of Dimensioning Overflow Basins in Combined System

(Richtlinien für die Bemersung U. Gestaltung von Regenentlastungsanlagen in Mischwasserkanälen ATV—Arbeitsblatt A-128 Entwurf: April 1990)

1. OBJECTIVE

Dimensioning of overflow basins (Regenüberlaufbecken) for the purpose of reducing, in combined system, the discharge into storm weirs (D.O.).

Temporarily stored water is dispatched to the treatment plant where it is treated as ordinary waste water.

In a general way, the criterion adopted for dimensioning consists of proceeding in such a manner that the pollution associated with run-off discharged in the exterior environment (= pollution discharged in D.O. + pollution discharged after treatment at downstream of the STEP) does not exceed the storm pollution discharged in the case of a separate system (see Sec. 2.7 below).

Computation is done on the basis of flow rate, volume etc., expressed in terms of annual mean.

This method is probably the only one of phenomenological type (not based on detailed modelling of the hydraulic and sedimentological process) that integrates the phenomena of pollution of rainstorm run-off.

Note: The symbols used for physical quantities (flow rate, concentration, etc.) are identical to those of the document ATV 128 [37].

2. DETERMINATION OF NECESSARY OVERALL STORAGE VOLUME

2.1 CHARACTERISATION OF CATCHMENT AREA SUPPLYING TREATMENT DEVICE

(i) height of annual precipitation: h_{NA} (mm).
(ii) surface of catchment region:

A_E = total surface area of catchment region

A_U = $VQ_R /(10 h_{NA, EFF})$ (ha)

= surface area activated from the point of view of run-off

VQ_R (m³) = volume of annual run-off

$h_{NA, EFF}$ (mm) = effective annual precipitation (= precipitation–losses, notably by infiltration)

VQ_R can be estimated by means of a model of rainfall-flow rate relation. It is generally assumed that A_U = covered surface area

(iii) Time of concentration of basin: t_F can be calculated by summing the times of water travel along the longest watercourse.

(iv) Group of mean slope NG_m.

The directive ATV conventionally recognises four distinct groups of slopes.

Group NGi (under B.V.i)	Mean slope
1	$J_G < 1\%$
2	$1\% \leq J_G < 4\%$
3	$4\% < J_G \leq 10\%$
4	$J_G > 10\%$

For a group of catchment areas

NG_m = $\Sigma(A_{E, i} \ NG_i) / \Sigma A_{E, i}$

where NG_i denotes the group of subbasin slopes.

2.2 CHARACTERISTIC FLOW RATES

(i) Flow rate (combined) admissible at the inlet to the treatment plant: Q_M = $2 Q_S + Q_F$, where Q_S and Q_F respectively denote flow rate of the dry season and parasitic water.

Remark: The directive does not specify whether Q_S is the value averaged over 24 h or over the period of heaviest precipitation (14 to 18 h).

(ii) Flow rate of the dry season averaged over 24 h: Q_{T24}

Q_{S24} (flow rate E_U) = $Q_{H24} + Q_{G24} + Q_{I24}$ (l/s)

where the subscripts H, G and I refer respectively to the habitat, crafts and industry. These flow rates constitute annual averages.

Q_{T24} = $Q_{S24} + Q_{F24}$ (l/s).

158

Q_{F24} denotes the flow rate of parasitic water.

In the absence of specific measures, Q_{G24} and Q_{I24} are assumed to lie between 0.5 and 1.2 l/s/ha (flow rates recorded on the surface A_U). Similarly, Q_{F24} may be assumed to lie between 0.03 and 0.15 l/s/ha (/surface area A_U) depending on the state of the sewerage system. Thus, one may note: $q_{T24} = Q_{T24}/A_U$.

(iii) Flow rate of the dry season averaged over daytime (duration x = 14 to 18 h)

$$Q_{SX} = \frac{24}{X} Q_{H24} + \frac{24}{a_G} \frac{365}{b_G} Q_{G24} + \frac{24}{a_1} \frac{365}{b_1} Q_{I24} \text{ (l/s)}$$

$$Q_{TX} = Q_{SX} + Q_{F24}$$

where $a_{G,1}$ = duration of workday (8 h) and $b_{G,1}$ = number of working days in a year.

(iv) Flow rate of rainstorm run-off entering the treatment device: Q_{R24} (ls) Q_{R24} is defined as the difference between Q_M and Q_{T24}.

Also, $q_R = Q_{R24}/A_U$ (l/s/ha).

Remark: If $Q_M = 2 Q_{SX} + Q_{F24}$ (see (i) above), $Q_{R24} = 2 Q_{SX} - Q_{S24}$.

(v) Mean flow rate (rainstorm run-off): Q_{RE}

Q_R = flow rate of discharge in the ensemble of D.O. + flow rate of rainstorm run-off diverted to the treatment device

$= [V_{QE} / (T_E 3.6)] + Q_{R24}$ (l/s)

where

V_{QE} (m³) = overall volume discharged annually in the D.O.

T_E (h) = annual duration of operation of the D.O.

Q_{RE} can be obtained approximately from the following equation:

$Q_{RE} = a_F (3 A_U + 3.2 Q_{R24})$ (l/s)

in which:

$a_F = 0.50 + 50/(t_F + 100)$ if $t_F < 30$ min

and $a_F = 0.885$ if $t_F > 30$ min

t_F = time of concentration (Sec. 2.1 (iii))

A_U = covered surface area (Sec. 2.1 (ii)).

Remark: The preceding equation is phenomenological, similar to formulas which, like rational formulas, determine a flow rate of run-off in terms of the mean intensity of rainfall during the time of concentration. On the other hand, one may ask why Q_{R24} appears on the right-hand side. Of course, the coefficients of the equation depend on local rainfall; they should therefore be modified if the method is applied to other regions.

2.3 Concentration of COD in effluents of the dry season

$$c_T = (Q_H \cdot c_H + Q_G \cdot c_G + Q_I \cdot c_I)/(Q_H + Q_G + Q_I) \text{ (mg/l)}$$

2.4 Ratio m in mixture

$$m = Q_{RE}/Q_{T24}$$

where

Q_{RE} is the annual mean of flow rate of rainstorm run-off (Sec. 2.2 (v)),
Q_{T24} the flow rate of the dry season averaged over 24 h (Sec. 2.2 (ii)).

2.5 Standard values of concentration of COD

(i) $c_C \cong 600$ mg/l (see Sec. 2.3)

c_C = COD in the rainstorm run-off (**before it mixes with the waste water in the combined system**) 107 mg/l

c_K = COD in the effluent of the treatment device
$\cong 70$ mg/l

(ii) Correction in c_T (standard) in terms of the actual value of c_T, annual precipitation and the phenomenon of deposition/resuspension in sewers of small slope:

$$c_B = 600 (a_C + a_H + a_A) \text{ (mg/l)}$$

where:
Correction due to actual value of c_T
$a_C = 1$ for actual $c_T = 600$ mg/l
$a_C = c_T/600$ for actual $c_T > 600$ mg/l.

Correction due to annual precipitation.
   The basic idea in this case is that if precipitations increase, the combined water will be discharged during longer periods towards the exterior environment. The corrective coefficient a_H thus simply expresses this increase:

$$a_H = \frac{h_{NA}}{800} - 1 \quad \text{for } 600 \leq h_{NA} \leq 1000 \text{ mm}$$
$$a_H = -0.25 \quad \text{for } h_{NA} < 600 \text{ mm}$$
$$a_H = +0.25 \quad \text{for } h_{NA} > 600 \text{ mm}$$

Correction for taking into account the phenomena of deposition/resuspension in sewers of small slope:

160

$$a_A = \left(\frac{Q_{TX}}{Q_{T24}}\right)^2 \left(\frac{2-t}{10}\right) \quad (a_A \geq 0)$$

where $t = 0.43 \, (q_{T24})^{0.45} \, [1 + 2 \, (NG_m - 1)]$

Remark: a_A thus lies between 0 and $0.2 \left(\dfrac{Q_{TX}}{Q_{T24}}\right)^2$ It can be verified that a_A decreases as NGm increases. On the other hand, a_A increases with $\dfrac{Q_{TX}}{Q_{T2}}$ and this behaviour is due to the fact that depositions take place mainly at night.

It is strange that the directive ATV attributes the phenomenon of deposition/ resuspension to waste water rather than rainstorm run-off.

2.6 Concentration (c_E) of COD in mixture of rainstorm run-off and waste water

From the relation:

$$c_E \, (Q_{RE} + Q_{T24}) = c_R \, Q_{RE} + c_B \, Q_{T24}$$

one obtains:

$$c_E = (m \, c_R + c_B)/(m + 1) \qquad \text{(mg/l)}$$

2.7 Rate E_o of admissible discharge in the D.O.

In accordance with the objective:

$$SF_E + SF_K \leq SF_R$$

where

SF_E (kg) = COD discharged annually in the group of storm weirs (kg),

SF_K (kg) = COD discharged annually in the effluent of the treatment plant (kg)

SF_R (kg) = COD annually washed out by run-off (kg)

= COD discharged in the exterior environment if the network is separate.

This equation can be transformed to:

$$VQ_R e_0 c_E + VQ_R(1 - e_0)c_K \leq VQ_R c_R$$

wherein VQ_R designates the volume of average annual run-off (see Sec. 2.1 (ii)), so that

$$e_O \ (\%) = [(c_R - c_K)/(c_E - c_K)] \times 100$$

When the medium of discharge is characterised by a mixture, ratio MNQ/Q_{SX} > 100 (MNQ is the mean flow rate of low water), e_O can be modified by a factor varying linearly from 1 ($MNQ/Q_{SX} = 100$) to 1.2 ($MNQ/Q_{SX} \geq 100$).

The effective rate of discharge e differs from e_O and the difference depends on the actual conditions of precipitations.

The directive ATV suggests the following expression for a rough estimate of e:

$$e = e_O + (h_{NA} - 800)/40 \ (\%)$$

2.8 Required overall storage volume

It is known that $V_S = V/A_U$ (m³/ha), V (m³) denoting the overall volume necessary for ensuring, on average, a treatment of the fraction $(1 - e_O)$ of run-off:

$$V_S = \frac{H_1}{e_O + 6} - H_2 \tag{1}$$

where H_1 $(4000 + 25 \ q_R)/(0.551 + q_R)$

and H_2 $(36.8 + 13.5 \ q_R)/(0.5 + q_R)$ (2)

q_R being expressed in l/s/ha (see Sec. 2.2 (iv)).

From formula (1) and expressions (2), V_S can be obtained directly in terms of e_O and q_R.

On the other hand, according to directive ATV, V_S should lie between the minimum value $3.60 + 3.84 \ q_R$ (detention time = at least 20 min to ensure conditions of deposition) and the minimum value 40 m³/ha (value estimated on the basis of technoeconomic considerations).

Also, the conditions for application of the above method are:

$$0.2 \leq q_R \leq 2 \ l/s/ha,$$
$$25 \leq e_O \leq 75\%.$$

3. OVERFLOW BASINS, DIMENSIONING INDIVIDUALLY

3.1 Method of distribution of overall volume calculated earlier, in terms of individual volumes

The method described in the preceding section is applicable successively to each

individual basin with the following modifications:
- Taking into consideration the catchment area situated upstream of the storage basin (A_U, t_F, etc.).
- Replacing Q_R (admissible flow rate of rainstorm run-off to the treatment plant) by Q_D (flow rate to the control station downstream of the basin), the latter being determined by means of the flow rate at the beginning of storage and overflow to the D.O.

Remark: The following point is not explicitly mentioned in the directive
- Verification of compatibility of calculated volume with minimum and maximum values mentioned above.

3.2 CONDITIONS OF APPLICATION

- q_R (upstream basin) $\leq 1.2\ q_R$ (treatment device);
- not more than 5 basins in a series, otherwise the method of distribution loses accuracy;
- not more than 5 D.O on the catchment area, which is a tributary of a storage basin.

3.3 PARTICULAR CASE OF STORAGE IN CHANNEL

When the overflow of such a basin is located near the downstream end of the structure, the required storage volume is increased by 50% with respect to that obtained by the preceding method. This precaution is related to the fact that the frequency of functioning of the downstream overflow should be reduced to limit the risk of entrainment of the most polluted water (bottom) towards the exterior environment under the action of currents.

3.4 MINIMAL DIMENSIONING OF BASINS

For reasons of construction, it is recommended that overflow basins of capacity less than 100 m³ (case of overflow downstream: *Durchlaufbecken*) or 50 m³ (case of basins not equipped with overflow device: *Fangbecken*) be not planned.

3.5 MINIMUM RATIO OF MIXTURE

One should verify for each overflow basin that the ratio m of the mixture remains ≥ 7 on average over a long period (case in which $c_T \leq 600$ mg/l).
If $c_T > 600$ mg/l, this limit is increased:

$$m \geq (c_T - 180)/60$$

3.6 Conditions of decantation
(Downstream overflow basin: *Durchlaufbecken*)

— Good conditions of decantation: rate of rise of water level in the basin ≤ 10 m/h.
— Safety against risk of resuspension of deposited matter: velocity of longitudinal flow in the basin < 0.05 m/s during a storm episode of intensity 30 l/s/ha.

3.7 Duration of evacuation

This should not exceed 10 to 15 hours

Remark: Although not stated in directive ATV, the reason is probably to avoid consolidation of deposited matter.

3.8 Case of retention basins of rainstorm run-off
(without D.O.) upstream (*Regenruckhaltebecken*)

It is assumed that if the rate of evacuation of these upstream retention basins is higher than 5 l/s per impermeable hectare of the catchment area, an extension of the works, the method of dimensioning of downstream overflow basins need not be modified.

But this does not hold if the rate of evacuation is lower than the above limit nor if it is higher than the admissible rate of flow (Q_R) of rainstorm run-off into the treatment works and another method of dimensioning of overflow basins (not described here) should be applied.

3.9 Circular overflow basins

The ATV directive states that at present there is no satisfactory method for dimensioning such basins.

ANNEXURE II

CASE STUDIES
(Meeting at Achères 19 May 1995)

Introduction to Meeting at
Achères and Nine Case Studies

While opening the meeting Mr. **Ténière-Buchot** recalled that storm weirs constitute a very important topic of discussion and cover not only the region Ile de France, but the entire country; he raised the problem of **financing** works whose objective is to reduce the discharge of urban wastes during the rainy season. He emphasised in particular the **need for specific studies** and for taking into account **landscaping and aesthetics**.

In the management of networks and their storm weirs, **innovation** should be given a place of importance. Researchers and persons associated with development have an important role to play in gathering existing information and finding effective solutions.

He noted that the meeting could not be meaningful without the work carried out by Mr. **Valiron** for the Agency and by the group of AGHTM headed by Mr. **Tabuchi**.

Lastly, he thanked SIAAP for giving support to the meeting by contributing two papers and providing reception and lunch for about 150 participants.

Before inviting the speakers, Mr. **Ténière-Buchot** requested Mr. **Valiron** and the persons in charge of large networks of the Ile de France to briefly present the results of studies and surveys carried out and the points meriting particular attention of the participants.

Mr. **Valiron** described the present state of progress in design and construction in the Parisian region and presented survey results. He reviewed the main types of existing storm weirs and illustrated with various examples the complexity of the management of sewerage networks of the Parisian agglomeration: complexity of the catchment area and collection system. Thereafter, he discussed differences in the management of each network, which are resolved by systematic exchange of data.

With respect to the report on the British experience circulated to the participants, Mr. **Valiron** showed through some examples the need for learning from it, as it was based on 20 years of work towards improvement of weirs.

Lastly, he presented a document titled **'Summary of knowledge regarding storm weirs'** edited by him. This paper focuses only on the Parisian region but proposes extension to other cases; he requested the participants to suggest any modifications or additions they would like to include. The main points of the case studies to be presented would, of course, be incorporated and constitute part of the collection on rainstorm run-off planned by the Agency and AGHTM.

On behalf of SIAAP, Mr. **Affholder** presented a brief history of the sewerage system of the Parisian agglomeration. He recalled that the first storm weir of the agglomeration, that at Clichy, dates back to 1895 and that weirs were initially constructed to provide protection against risk of exceeding the capacity of a combined network.

It is only recently that attention has focused on optimisation of the functioning of storm weirs and limiting the pollution discharged directly into the natural environment. Efforts to reduce the pollutant load are presently directed towards reduction of pollution during the dry season.

Abandonment of Achères V has necessitated consideration at the **global level of the problems of pollution of both the dry and wet seasons** on the Parisian agglomeration. The objective is to **limit fish mortality in the Seine River.** This objective requires determination of the maximum admissible pollution load.

To achieve this objective further requires improving solutions combining **optimal utilisation of the capacity of treatment, transport and storage.**

On behalf of the **city of Paris**, Mr. Constant described the various **measurement programmes carried out on the Parisian network since the 1980s.** He recalled that the survey done in 1988-1989 on the Bièvre and Bas sewers had demonstrated the preponderant role of the Bièvre and Alma weirs. Subsequently, from June 1993 to September 1994, 16 of the large weirs were equipped with hydraulic measurement devices.

As a follow-up of a survey initiated in 1994, the work of equipping the major installations of the network with measurement devices will continue in 1995; 80 measurement sites will be established and equipped with limnimeters, flowmeters and rain gauges.

The studies and measurements showed that the largest overflows occur in the eastern part of the capital; the weirs involved would therefore be treated on a priority basis.

On behalf of **Hauts de Seine,** Mr. **Millet** expressed anxiety regarding the **complexity of the problem** by posing the question: 'What shall we do about our water?' He mentioned the efforts made towards **optimisation of the storage capacity** of the network at the mouth of the Boulogne. According to the simulations carried out on the rainfall of 4 December 1992, the volume of overflow could be reduced by half.

Mme. **Cogez**, representing **Seine St Denis,** recalled that the **Seine St Denis** had been carrying out various experiments (static vortex separator, expandable barrages etc.) for several years. Today, St Denis has a sound knowledge of management in real time as well as good control of the hydraulics on its network.

She stated that reduction of the **pollution generated by the urban discharge during the rainy season and the battle against inundation are complementary.**

At present, the main lacuna arises from inadequate knowledge of the impact of overflow on the natural environment. Among these impacts, aesthetics is an important aspect and must be taken into account.

One of the main difficulties of **management in real time** arises from the fact that the storm event is not known a priori. The manager of the network has to anticipate the disorders which could eventually occur due to rainfall, in order to make the necessary decisions. He does not have the authority to undertake large risks. **But how to optimise anticipation?**

The simulation of real rainfall, following various scenarios of management, may provide a posteriori useful information which could be compared with the approximate real situation of flow rate. One may use this information to evolve, step by step, a more and more refined strategy of action, depending on the time available and one's experience.

On behalf of **Val de Marne,** Mr. **Darras** recalled the constraints faced by the district of Val de Marne. Both on the Seine and Marne the intake of water for the supply of **potable water imposes an onerous task on the manager of the network in terms of maintenance of water quality.** It requires **overall management and technical co-operation** with the community to acquire better knowledge and control of what happens upstream.

Reduction of the impact of urban discharge during the rainy season requires more **profound knowledge and quantification of rainfall.**

After a vote of thanks to the speakers, Mr. Ténière-Buchot requested various authors to present their papers.

At the end of each presentation, the reader will find a few of the questions posed during the discussion; these are printed in italics.

Impact on the Seine River of the Discharge from Storm Weirs. Results Obtained by the Group 'Urban Catchment Areas' as Part of the PIREN-Seine Programme

Jean-Marie Mouchel

(CERGRENE Ecole Nationale des Ponts et Chaussées,
La Courtine, 93167 NOISY-LE-GRAND Cedex)

The results presented here were obtained by the 'Urban Catchment Areas' group of the PIREN-Seine programme. Names of all the participants in the programme would take too long to mention but I thank each for his/her contribution, in particular my colleagues from LABAM (University of Créteil), CERGRENE, CIG (School of Mining), LGA (University of Paris 6) and GMMA (University of Brussels). I am also grateful to the institutions which provided financial and technical assistance during this five-year study

INTRODUCTION

The work concentrated primarily on a study of the impacts of urban discharge during the rainy season in the Parisian agglomeration and analysis of the mechanisms of transfer and action of the pollution discharged. Efforts were also made to model the aforesaid processes by conceptualising the biological mechanisms discussed by the team 'Axe Fluvial' (**Servais** et al., 1993 [93]), **Garnier** et al., 1993 [92], **Billen** et al., 1993 and constructing uni- or bi-dimensional working hydraulic models (**Even and Poulin,** 1993 [88]), **Simon,** 1995 [97]). Although the software for simulations prepared by us can be used as is, the group 'Urban Catchment Areas' of PIREN-Seine is constantly improving it, notably the programme of measurements planned for 1995.

The impacts of the urban discharge of the rainy season are multiple and the underlying principle has been discussed in various articles, in particular by **Chocat** [84]. Broadly speaking, pollution can be divided into several types likely to affect the environment at different time scales. In the case of the Seine, the impact of organic matter around and downstream of the Parisian agglomeration is particularly evident and is accompanied by the phenomenon of deoxygenation, which can now be accurately measured and is sometimes disastrous for fish life. The concentration of ammonium downstream of urban discharge during the rainy season can be very high in the Seine (of the order of several tens mg/l, the value peaking immediately downstream of the point of discharge) but the concentration of ammonia directly toxic to fish is much smaller because the pH of the Seine seldom exceeds 8. At pH 8 and 25°C only 5% of the ammonium is in the form of ammonia (**Gammeter and Frutiger, 1990 [91]**); at lower pH and/or temperature the quantum is less. The value 0.5 mg/l is approximately the lethal threshold for rainbow trout with an exposure time of 24 hours (**Garric** et al., 1990 [93]). Thus ammonium would not appear to be the major factor for fish mortality; although its concentration seems to be high, it is lower, by an order of magnitude, than the lethal threshold. The presence of flotsam in the river after storms is aesthetically unpleasant, so its removal has been organised by designated services. On the other hand, pollution due to micropollutants and bacteriological pollution (in terms of pathogenic bacteria) are not well known although theoretically such could be significant. Lastly, let us mention pollution due to nutrients and suspended matter; apparently this type of pollution is not significant in the Seine during the rainy season, but relatively large quantities are present in the urban discharge of the dry season and in agricultural discharge. The latter situation can be rectified with proper treatment.

The 'Urban Catchment Areas' group of PIREN-Seine has concentrated on the study of deoxygenation and micropollutants, in particular metallic, in the Seine after the discharge of the rainy season. Although their number can be rather large, de facto pathogenic organisms do not reach an alarming proportion in the Seine in the interior of the Parisian agglomeration due to lack of swimming resorts. Nonetheless, this aspect should be studied since swimming resorts might be restored in the region upstream of the agglomeration and, truth to tell, unauthorised swimming sites and nautical sports do exist. This problem has yet to be studied by the 'Urban Catchment Areas' group due to lack of funds, certainly not lack of concern!

The first step towards well-controlled modelling consists of striking balances in order to analyse whether the processes of transformation are effective at the scale of the site under study or their efficiency can be taken as constant; the principal variables on which this efficiency is dependent have also to be ascertained. Determination of a balance is not easy. Correct results require a very large number of measurements, rarely practicable. One has always to take recourse to simplifying hypotheses (in principle, rational) such as transverse homogeneity in a river at a sufficient distance from the point of discharge, or temporal homogeneity of the situation upstream of the point of discharge, which enables attribution of the changes observed downstream to discharge only. When the hydraulic characteristics are not well known or a general problem of transport exists, tracers are useful in resolving the difficulties. But some standard cases notwithstanding, employment of tracers also depends on hypotheses that are similar when obtaining balances (spatial or temporal homogeneity of some of the sources).

1. DEOXYGENATION

We worked in the Suresnes-Chatou pond situated in the interior of the Parisian agglomeration, downstream of Paris but upstream of the treatment works of Achères whose discharge would have strongly perturbed the balances. There are two main islands in the pond: St Denis at the level of the point of discharge of the Briche and Chatou in the downstream part of the pond. The water quality in the pond was intensively studied during two successive summers—1991 and 1992. We were primarily interested in determining the impact of the Clichy discharge whose water quality was known, thanks to a recent study by SIAAP (**Paffoni,** 1992, 1994) and in which the volume of discharge could be determined by the SCORE system (**Le Scoarnec and Bouvet,** 1990).

Figure 1 shows the principal characteristics of water discharged at Clichy during the rainy season. BOD_5 is the best indicator of degradable organic matter likely to cause oxygen deficiency in the Seine. The concentration of BOD_5 in the discharge is only slightly higher than that during the dry season. Its variation is sufficiently small because the typical deviation of the concentration is only

	Dry Season	Overflow
SS (mg l^{-1})	144 (73-228)	279 ± 89
BOD_5 (mg l^{-1})	76 (36-101)	88 ± 28
COD (mg l^{-1})	252 (103-400)	315 ± 108
Conductivity (µS cm^{-1})	750 ?	493 ± 120

Fig. 1: Concentration of pollution at Clichy. The concentration of the dry season is the mean concentration, weighted by the flow rate. Numbers in parentheses are minimum and maximum values. The concentration of the rainy season (in fact, during the overflow of the Clichy weir) is first averaged over the events, proportional to the flow rate. Typical deviation given in the Table characterises the distribution of the mean concentration over the events. These results are taken from Paffoni (1992).

30% of the mean value. The data obtained by SIAAP (**Paffoni**, 1992) for 22 events show the effect of dilution only for the largest overflow, close to 1,200,000 m³ in volume. In view of the variation of storm events, larger differences were expected. The data compiled for small separate basins by Saget (1994) give much more variable mean values of BOD_5 during the event, the typical deviation being larger than the mean value while the combined drains of Seine StDenis show less variable BOD_5. The large size of the catchment area and the presence of waste water mixed with the run-off contribute to the effect of smoothening the concentration.

The mean value of concentration of BOD_5 constitutes in itself a measure of the water quality of overflow at Clichy. We tried, but unsuccessfully, to improve this prediction by taking into account the quantity of run-off and various simplified representations of the mixture of waste water and run-off (**Mouchel**, 1993 [95]). These simplified models enabled improvement of prediction with respect to conductivity or ammonium, which correlated strongly with the quantity of waste water, but there was no improvement in prediction of BOD_5. The matter deposited on the urban surfaces and in the interior of the sewerage network also contributed to the BOD_5 and offset the reduction realised by dilution of less contaminated waste water.

Downstream of the overflow of the combined network of the agglomeration one could observe at all treatment (Suresnes, Colombes, Chatou and Bougival) a marked depression in the curves of dissolved oxygen (as the polluted water flowed by). In some cases, the concentration of dissolved oxygen was zero.

We were particularly interested in the works at Chatou whose results can be directly used because in this case the deficiency is more marked than that of Suresnes and the effect of Clichy and Briche can be better specified in a large number of cases. During the summer of 1991 the effects of 8 of 9 overflows at Clichy could be quantified. During the summer of 1992, when the rainfall was more intense, of 19 overflows at Clichy the effects were measured on 13 occasions; in other words, 21 events in two years could be utilised.

For each of these events, we computed the depression in the curves of dissolved oxygen as it appeared on the recorder (Fig. 2). Of course, the surface could have been accurately calculated from the mean data at intervals of 15 min but this method was marred by two sources of error: (i) the exact instance of the start and finish of the passage of water could not be correctly detected and (ii) it was impossible to accurately ascertain what the curves of dissolved oxygen would have been in the absence of overflow at Clichy. We assumed rectilinear behaviour between the start and finish of the passage of polluted water.

The surface area calculated above should be multiplied by the flow rate in the Seine at the instant of passage of polluted water, so that the number of tons by which oxygen is deficient can be estimated. A correction was applied that

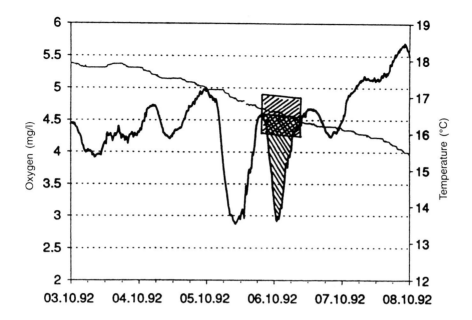

Fig. 2: Concentration of oxygen and temperature measured at Chatou. The two successive troughs are attributed to the discharges of Briche and Clichy. The hatched part shows the affected area which was calculated and for which the potential error was estimated to be 3 mg/l.

took into account the overestimation of the deficit due to poor lateral homogeneity of concentration upstream of the island of Chatou. The coefficient of lateral dispersion was assumed to lie between 0.04 and 0.07 m²/s (**Bujon,** **1983** [83]; **Fisher** et al., 1990 [90]). Elementary simulations of dispersion showed that the overestimation was 20 to 30% for a flow rate varying from 80 to 120 m³/s.

The quantity of oxygen passing through Chatou in the Seine after an overflow is given by:

$$Q = Q^* - D_i - C_{ex} + R_{ex} \qquad (1)$$

where Q^* is the quantity which would have flowed in the Seine in the absence of deficit, D_i the initial deficit due to inflow of deoxygenated water (water discharged), C_{ex} the overconsumption of dissolved oxygen due to discharge and R_{ex} the excess of reaeration caused by the lower than normal concentration of oxygen. The quantity $(Q - Q^*)$ is the deficit calculated from the measurements. It was estimated by assuming that the concentration of dissolved oxygen in the discharged water was 1 mg/l while R_{ex} was determined using a reoxygenation coefficient equal to 0.2 per day and assuming an almost linear variation of concentraton between the point of discharge and Chatou. Tables given in the

175

appendix (see **Mouchel**, 1993 [95]) illustrate these results and give the respective weightage of different terms.

Several factors complicate the relationship between the observed deficit (C_{ex}) and BOD_5. In particular, the water temperature and time of water transit influence the quantity of organic matter which could be degraded before arrival at Chatou. We have taken these effects into account according to the usual procedure based on the concept of Streeter and Phelps, by using a first-order kinetics with coefficient of degradation 0.6 per day at 20°C and coefficient of variation of degradation with temperature equal to 1.05 per °C. Apparently, this model is unable to explain the observed deficits because the ratio of the observed value to the theoretical value varied between 0.05 and 3.9.

These results are shown in Fig. 3 in terms of volume of overflow at Clichy. It seems that the larger the discharge, the smaller the deficits with respect to the value obtained from the simple model of degradation; contrarily, the smaller the

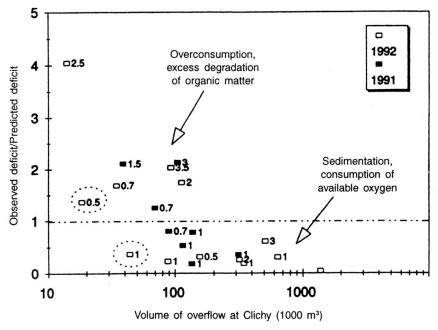

Numbers alongside the symbols indicate the abundance of phytoplankton in the Seine.

Fig. 3: Quality of simulation obtained by a simple model of Streeter and Phelps. Observed deficits (quantity by which oxygen falls short) are larger than the predictions for small overflow and smaller for large overflow. At equal overflow, the presence of phytoplankton increases the deficit. The two points encircled indicate the series of measurements potentially disturbed by installation of devices for reoxygenation.

176

discharge, the larger the deficits. By changing the temperature coefficient from 1.02 to 1.08 (instead of 1.05), no significant change in the respective position of points in the Figure was observed (change hardly of the order of 10%); of course, the temperature in the river is not uniform.

One explanation for the failure of the Streeter and Phelps model is the sedimentation of organic matter in the course of its trajectory in the Seine, the deposited matter consuming quantitatively less oxygen and causing an eventual deficit in the water that follows. The larger the discharge, the more pronounced this effect. There are two reasons for this trend: first, a heavy rainfall in a combined network generates a pollution which is easier to decant than that generated by light rainfall (**Chebbo**, 1992); secondly, with a larger discharge the maximal capacity of transport of suspensions in the river (in the direction of Velikanov) is attained and hence an increment in matter deposited.

Although the above explanation partially justifies very small deficits for the heaviest rainfall, it does not justify the excess deficits recorded for the lightest rainfall. Thus the concept of Streeter and Phelps fails to describe this type of short-duration event. The process of oxygen consumption by bacteria cannot be described by first-order kinetics; in the case of small overflow there is an overconsumption in the first instant, a day after the overflow. It is due to the combined dynamics of bacteria and organic matter present in the Seine and that brought in by the discharge. In particular, the organic matter of planktonic origin plays a special role.

Although most of the models which simulate the degradation of organic matter in the natural environment use first-order kinetics, it is important to bear in mind that the process is, above all, biological and thus depends, in the first instance, on the quantity and activity of the heterogeneous biomass present in the environment. A second important factor is the quality of the organic matter, since some molecules such as humic acids do not readily biodegrade.

Heterotrophic bacteria have an efficiency of the order of 30% in the Seine: for every two molecules of carbon which are degraded (transformed to CO_2) one is incorporated in the heterotroph biomass. Thus the 'heterotroph biomass' plays a double role: on the one hand it is the active component whose presence controls the rate of degradation of organic matter and, on the other, it is one of the quantitatively larger components of organic carbon through which a large part of carbon transits.

The maximum rate of growth of heterotrophic bacteria, in conditions of culture over directly assimilable substratum, is 2 h^{-1} (**Edeline**, 1979 [85]). In the Seine, the observed maximum rate of growth is 0.2 h^{-1} for robust bacteria but in waste waters 0.04 h^{-1} for frail bacteria, more naturally present in the Seine (**Servais** et al., 1993 [96]). Thus the time of doubling of the bacterial population under optimal conditions is at the least 3.5 hours for robust bacteria and 18 hours for the frailer. If too few bacteria are brought in by the overflow of a combined network, a large delay may be observed before the process of deoxygenation develops.

The process of degradation given in the HSB model is presented in Fig. 4, taken from **Servais** et al. 1993 [96]. It illustrates the importance of the quantum of heterotroph biomass as it influences almost all the processes. This scheme forms the basis of models used in the PIREN-Seine programme.

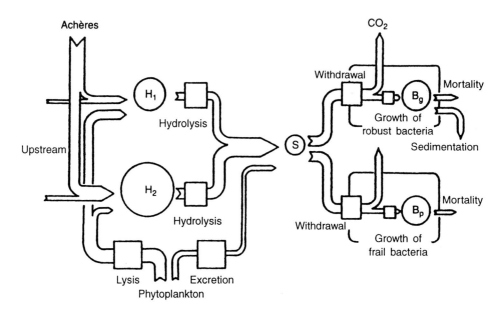

Fig. 4: *Diagram of the HBS model as adopted by PIREN-Seine models (from Servais et al., 1993). The bacterial compartment is both the promoter of biodegradation and one of the compartments which constitute the pool of organic matter.*

The urban discharge of the rainy season as well as the development of its effects in the natural environment are strongly transiting phenomena, both from the hydraulic point of view and that of water quality.

In extreme cases, the maximum rate of overflow from a single point of discharge might be larger than the flow rate of the Seine. The multiple points of discharge during the rainy season in the interior of the Parisian agglomeration could thus cause a flow rate not readily predictable. Although the waves disappear rapidly after the discharge, significant water movement is often observed between the instant of deficits recorded downstream— at Chatou in this case—and the theoretical instant of passage calculated on the basis of the mean flow rate of the Seine during the period covering the event.

From the quality point of view, it has been illustrated that use of a first-order model of degradation completely ignores the kinetics of growth of heterotrophic flora; also, it is very difficult to assimilate in this model the first few hours following overflow, which can be crucial to development of minimum oxygen.

The PROSE model, developed by the School of Mining, is particularly adapted to the simulation of urban discharge of the rainy season. It is based on hydraulic characteristics of St Venant and is thus capable of simulating flows around the islands. It can also simulate the effect of movement of barrages which are very important during storm events because of the requirement to maintain the water body of the Seine on a fixed side. This further complicates the movement of water in the Seine during events of heavy rainfall.

From the biological point of view, PROSE is capable of simulating the activities of heterotrophic bacteria and phytoplankton whose activity can be strongly inhibited during rainy periods. Some simulations were presented by **Even** et al., 1993 [89]. Other researchers are trying to simulate the summer of 1991 by using both data on oxygen in the Seine and measurements of quality of discharge at Clichy. The passage of water deficient in oxygen due to discharge at Clichy has been successfully simulated at Chatou by taking into account the hydraulic contribution of other discharges simulated by means of FLUPOL, using rainfall data estimated by radar. The trough of oxygen is well demarcated but slightly displaced upwards (which is evidence of inadequate simulation of the state of the Seine during the dry season).

Ideal operation of PROSE requires a better knowledge of discharge than that obtained from usual measurements; also, the results of models of run-off in the networks should be calibrated with the measurements. In particular, it is necessary to know the quantity and size of the bacteria discharged as well as the quantity of particulate and dissolved organic carbon. We used the correlation between BOD_5 and these unknown quantities determined in water arriving at Achères. Specific work is in progress for refining these correlations for water of the rainy season in combined networks.

The latest improvement that the 'Urban Catchment Areas' group is trying to incorporate in PROSE concerns the transport of suspended matter and organic matter or bacteria contained therein. Preliminary work (**Tangerino**, 1994 [98]) has helped us in differentiating the velocity of fall of particles present in the Marne during the dry and rainy seasons (Fig. 5). The V_{50} is of the order of 1 m/h during the period of low flow rate and the dry season in the Marne—it attains a value of 2 m/h after an overflow—and at the interior of a water mass visibly affected by the overflow. It may be noted that the slope of the curve changes slightly below V_{50}, which enables prediction of important properties of particles whose velocity of fall is small (by extrapolating the curve of the dry season in the domain of small velocity of fall, one obtains about 25% of the particles whose velocity of fall is less than 1 m/h). On the other hand, bacteria do not seem to fall while organic carbon is entrained with the particles.

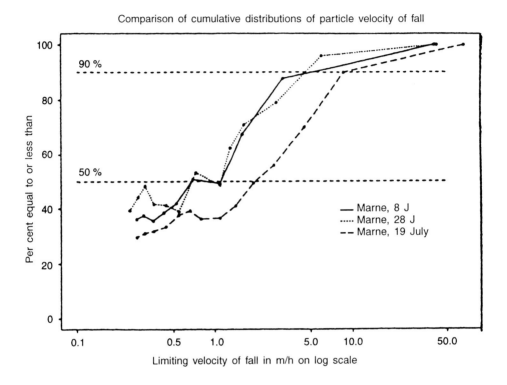

Fig. 5: *Distribution of velocity of fall obtained in the Marne (at Perreux) during dry and rainy seasons. On July 19, after a heavy storm, the velocity of fall is markedly higher (from Tangerino, 1994).*

2. HEAVY METALS

The 'Urban Catchment Areas' group of PIREN-Seine devoted a significant part of their effort to the study of the fate of metals in the Seine after storm overflows. The fate of metals associated with suspended matter is now well known. Further effort has to be devoted to dissolved metals in order to accurately ascertain the level of concentration attained in the Seine during the rainy season and an analysis done of whether the exchange between dissolved and particulate forms modifies the relative value significantly.

Alexandrine Estèbe and Anne-Laure Bussy of LABAM found the following trends:

— very significant sedimentation of particulate metals after overflow,
— partial release of settled metals towards the water column,
— marked continual resuspension of deposited matter during the first flood.

The particulate metals were monitored for 18 months in the Seine at Chatou using suspended traps. These devices enable continual collection of part of the suspended matter in transit in the Seine. The results obtained by **Estèbe** et al., 1993 [87] are shown in Fig. 6. The concentration of dissolved metals is high during periods of low flow rate and decreases rapidly during the period of flood; the concentration of iron or MVS, generally representing the most reactive fine fraction, undergoes small variation. Thus, it is a matter of a real change in the quantity of suspended matter and not a simple phenomenon of dilution by the particles which are large in size and therefore poor in all metals. But this very general description is inadequate, only a detailed analysis of results will demonstrate other very important aspects.

During the period of low flow rate, the peaks of concentration appear at Chatou after large discharges of the rainy season. They are a trace of the passage of suspended matter originating from the urban discharge of the rainy

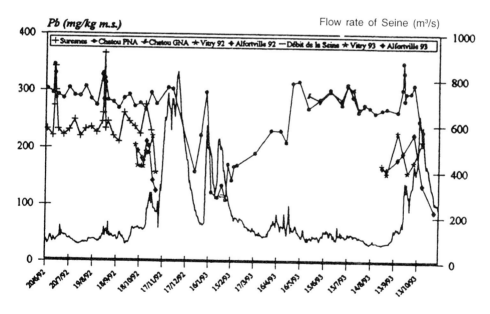

Fig. 6: *Quantum of lead in the suspensions of the Seine at the crossing of the Parisian agglomeration. Small peaks appear at low water after the main urban discharge of the rainy period. The lowest quantum occurs during floods but the reduction of concentration at Chatou occurs well after the first flood.*

season. After the rains, we increased the frequency of sampling to attain a temporal variation on a daily basis; this helped in defining the effect of passage of different quantities of water. Thus knowledge of the flow rate of solids in the Seine, quantity of SS discharged by Clichy (obtained from the data of **Paffoni**, 1992 and the data on flow rate provided by SCORE) helped us to estimate that about 50% of the SS originating from Clichy would have passed above the trap at Chatou during the day corresponding to the maximum concentration. A computation of dilution based on (i) the concentration of metals in the traps during the dry period and at the peak moment and (ii) the quantum of metals in the suspended matter originating from Clichy, enabled us to determine the percentages of matter originating at Clichy and collected in the traps; it was only about 5% (**Estèbe** et al., 1992 [86]). These results suggest that 90% of the SS originating at Clichy did not pass through Chatou at the moment of passage of the water mass originating at Clichy.

Of course, this conclusion is based on the hypothesis of a good representativity of traps. A more accurate statement would be '90% of the SS originating at Clichy and collected by traps does not pass at Chatou'. At present, we have only incomplete information on the actual representativity of traps in all situations regarding flow rate. Comparison among quanta of metals in the SS of the Seine collected by filters or traps was made in August 1993 with a flow rate markedly larger than 100 m³/s.

Figure 7 shows that the variability of measurements on filters is higher than that for traps; this is probably due to the integrating effect of traps. If the high value of the zone were excluded, the quantum would be higher in the trap than on the filters, especially of lead and copper. Apart from the mode of drawing samples, the chemical treatments employed are a source of significant bias, which might partly explain the differences observed. One would expect a smaller quantum in the trap because of a coarser particle grain size necessarily settling; but this does not actually happen. Although other points of comparison are needed to ensure good representativity of traps, the potential errors in concentrations leave no scope for doubting the results regarding sedimentation in the Seine of matter originating at Clichy. Zinc supports the argument strongly because in view of the mean level observed in SS at Clichy (**Estèbe** et al., 1992 [86]) in the absence of sedimentation, the concentration in the traps would be higher than 4000 ppm instead of the observed value of about 900 ppm at peaks and 750 ppm during the dry period.

In conclusion, in the absence of certainty with respect to the representativity of traps, we hope that the value 90% will be corrected in future. However, it can be asserted that a very large fraction of particulate metals (in particular, zinc) got deposited before reaching Chatou in the course of the two events analysed— the two most important events of the summer of 1992.

	Zn (ppm)	Pb (ppm)	Cu (ppm)	Cd (ppm)
Filters	756 ± 463 559* ± 163	162 ± 37	158 ± 35	4.3 ± 2.0
Traps	620 – 609	277 – 275	242 – 234	4.9 - 4.5

Fig. 7: *Concentrations on filters are taken from Huang and Mouchel (1994)*
[94]. Five samples were analysed at Chatal on 4 and 5 August 1993. One of
the values of Zn was very high (1550 ppm) and was excluded from the number
marked with an asterisk.*
The quanta on the traps are taken from Estèbe et al. (1992).
The two values correspond to two traps of slightly different design.
The traps were installed on 29 July and removed on 5 August, 1993.

This conclusion can be cross-checked with other information based on determination of the flux of particulate metals due to urban discharge of the rainy period transiting in the Seine. It is a matter of extrapolation based on a very simple hypothesis formulated for obtaining the order of magnitude of flux. For the period of low water in 1992 (June to October) we estimated the quantity of SS discharged at Clichy by using the mean concentrations measured in 1990-1991 (**Paffoni**, 1992) and the volume discharged by means of the SCORE system. This flux was subsequently multiplied by a factor of 6 for computing the discharge of the entire system of the agglomeration. This factor was taken from simulations carried out with FLUPOL for the entire agglomeration (**Dégardin**, 1991 [44]) so that the order of magnitude of different discharges could be compared. Lastly, we used the mean concentrations of SS at Clichy and La Briche (**Estèbe** et al., 1992) to calculate the flux of metals. In view of the methods employed, these results should be considered only orders of magnitude. The flux passing through the Seine was determined from the mean concentrations of metals measured in the traps, mean concentration of SS at low water and flow rate of the Seine at Austerlitz; the results are tabulated in Fig. 8.

It can be seen from Fig. 8 that the flux of zinc generated by the urban discharge of the rainy season is considerably larger than that passing through Chatou during the period under consideration. Although the estimates are rough, this Table shows that a significant quantity of zinc contributed by UDRS disappears. Although not very obvious, the same phenomenon occurs in the case of copper and lead. It is only in the case of cadmium that the quantity is unaltered, suggesting the existence of another significant source in the interior of the agglomeration. It should be noted that the excess of particulate zinc, copper and lead observed at Chatou vis-à-vis Vitry or St Maurice, could be due to a continuous suspension of matter contained in UDRS.

	Cd	Cu	Pb	Zn
Urban discharge of the rainy season (UDRS)	0.05	4.8	5.1	40
Chatou	0.09	4.1	4.7	13
Suresnes	0.05	4.0	3.9	8.7
Vitry and Saint Maurice	0.03	4.1	2.9	6.5

Fig. 8: *Estimated values of metal flux (in tons) during the period*
June to October 1992.

Lastly, this phenomenon of sedimentation may be further confirmed by the resuspension of matter deposited in the course of low water at the instant of the first flood.

During the summer of 1994 the flow rate was quite high (of the order of 150 m^3/s). Then with the first flood of 5 September, the flow rate attained the maximum value 300 m^3/s on 20 September, returning to the value 150 m^3/s on 27 October. It can be seen from the following figures that the series of samples collected during flood rise differ markedly from the series collected during flood abatement; in addition to the marked difference in concentration, the ratio Zn/Pb is 2.9 during the rise and 2.2 during abatement. This ratio should be compared with the concentrations measured in the summer of 1992 at Chatou (Zn/Pb = 2.8) and at Suresnes (Zn/Pb = 2.3); these figures are taken from **Estèbe** et al., 1992 [86] as well as measurements of the sediments of the Suresnes-Chatou pond carried out in the summer of 1994 (Zn/Pb = 2.8; from **Estèbe** et al., 1994). It may be noted that at Chatou, the ratio Zn/Pb decreased to 2.0 in the winter of 1992-1993 and the ratio obtained at Samois (near Fontainebleau) was 1.9. In the discharge of Clichy and La Briche this ratio attained the values of 10 and 5 respectively.

The change in ratio from upstream to downstream as well as its level in the UDRS indicates that the origin of SS in the agglomeration is the significant factor. During the rise in the September 1994 flood we observed the passage of very large quantities of SS (significantly in excess of 50 mg/l) with the Zn/Pb ratio similar to that observed during low water at Chatou and in the pond sediments but markedly higher than the ratio observed in the SS of Suresnes during low water. During flood abatement the concentration of SS was appreciatively lower while the Zn/Pb ratio remained close to that observed at Suresnes. We interpret these results in terms of the origin of sedimented matter

184

in the interior of the Suresnes-Chatou pond eroded during flood rise; erosion was much weaker during flood abatement because the erodable quantity was small or already exhausted and therefore the SS that passed through Chatou had to have come from upstream.

CONCLUSION

Let us now summarise the important points of this brief presentation.

The intensity of deoxygenation was compared to that which could be predicted from a Streeter and Phelps model. It appears that this model underestimates the deficits for small discharges and overestimates them for large discharges. Also, it would seem that for the same discharge the presence of phytoplankton in the river causes larger deficits.

These observations lead to several perspective conclusions. On the one hand, Fig. 3 illustrates that the deficits can be used for correcting predictions made from the Streeter and Phelps model provided the hydrological and climatic conditions used for sketching the diagram are comparable to the simulated conditions. On the other hand, for better prediction one requires a better understanding of the mechanism causing such differences between the predicted and actual values. In this vein of thought, we mention two trends worthy of follow-through.

These trends are related to the question of sedimentation in the Seine and the interaction betweeen the matter deposited and the water column. Our research on deficit of dissolved oxygen and quantum of metal demonstrates the importance of this question. These trends are also related to the study of overflow characteristics of combined networks by determining the quantity of heterotrophic bacteria carried by the discharge and the biodegradability of the organic matter (it can become more resistant to treatment in the discharge mobilising the erodable stock in the network or on the roadside). They also cover the overconsumption of oxygen in the presence of phytoplankton (which itself is due to the presence of additional organic matter, i.e., planktonic cells) which possesses special characteristics and is degraded in the course of these events.

DISCUSSION

Mr. Valiron: *I believe that storage in the Seine is an important parameter.* **But how to make use of this storage which modifies the process of self-purification?**

Would not some modification in the management of barrages which are small in terms of water level, improve the situation?

The roadways appear to be responsible for large quantities of **heavy metals.** *If so, it is necessary to make special provisions with respect to* **roadway management** *and to find the* **means for financing them.** *Why not charge the persons who are the source of this pollution, i.e., those who use motored vehicles?*

Mr. Mouchel: **The source of production of Zn and Pb is still not well known.** *The persons responsible for the offence should be brought to book.*

Mr. Tabuchi: *What is the* **influence of sedimentation on the consumption of dissolved oxygen** *(overall oxygen balance)?*

Mr. Mouchel: *The quantity of oxygen consumed by the bottom deposits is quite large. However, the Seine is a deep river and* **a consumption of 2 to 3 g/m³/day cannot entrain a significant effect** *on the total quantity of water.*

CASE NO. 2

Parisian Storm Weirs

A. Constant (City of Paris)

1. PRESENT SITUATION

1.1 SEWERAGE NETWORK OF PARIS

The sewerage network of Paris as it exists today was designed and most of it constructed during the second half of the nineteenth century on the initiative of Eugène Belgrand, the then director of 'Water and Sewers of Paris'. It covers a surface area of about 8500 hectares and serves a residential population of about 2.1 million plus about 1 million suburbanites who commute to Paris for work.

The network carries daily during the dry period 1.3 million m³ waste water or an average of 15 m³/s. It is a **combined** network and collects both rainstorm run-off and waste water; it also has some special features:

• It operates essentially on **gravity;** only five pumping stations are installed which lift the water from the 12th, 13th and 16th zones situated at a lower level.

• All works are open to inspection; the total length of the works is 2200 km, of which 170 km form sewers and drains, and 600 km auxiliary works (special connections, inlets, inspection galleries).

Apart from the sewers for waste water and run-off of Paris and their transport up to the drains from where they can be carried to the Achères treatment works, the network carries the effluents of suburbanites residing upstream of Paris through sewers linked to the same treatment works.

Protection of the network against overload and inundation during the period of heavy rainfall is provided by **43 storm weirs** discharging the excess water directly into the Seine and playing the role of safety valves.

1.2 PROTECTION OF NETWORK AGAINST SEINE FLOODS

During a flood the water from the river can invade the sewerage network through the storm weirs. To preclude this risk of inundation, the inlets to the Seine are fitted with watertight gates. In this situation protection of works against the extra load due to rain cannot be ensured by the weirs; two pumps are temporarily installed to evacuate the excess water into the Seine.

The sewerage service has prepared a manual of duties describing exactly the steps to be taken as and when the network is overloaded.

1.3 POLLUTION GENERATED BY OVERFLOW

The weirs fulfil the task of protecting the network against inundation but they also discharge a large quantity of pollution into the Seine during the rainy season. The thresholds of weirs, mostly fixed, are indeed adjusted to a low level to prevent excessive loading, even in the case of an unusual storm. For this reason, they cause overflow as soon as the rainfall exceeds a certain limit. A very large volume of combined water is thus discharged in the case of heavy rainfall, a volume which can reach a million m^3 in a storm of the decennial type. The consequences for the Seine are disastrous, especially in summer when the flow rate can be less than 100 m^3/s and the temperature rises above 25°C. Fish mortality was observed in June 1990, May 1992 and July 1994.

It may also be pointed out that networks of the suburbs upstream and downstream of Paris are also equipped with numerous weirs and so the polluting effects are cumulative.

1.4 CONTROL OF WEIRS

At many locations, the Section of Parisian Sewerage (SAP) has carried out measurements to determine the volume of overflow as well as water quality. A programme of measurements conducted for one year in 1988-1989 on the catchment area of Bièvre and Bas sewers enabled a complete balance of overflow on the left bank.

More recently, a programme of measurements was conducted on 16 weirs, chosen from among the largest, between June 1993 and September 1994; special attention was paid to measurement of discharge during the summers of 1993 and 1994. From a simplified hydraulic model it was estimated that these 16 weirs represent on average 80% of the discharge from the Parisian weirs.

• **In the summer of 1993** the rainfall was below the average (175 mm from June to September versus 200 mm in a normal year). The total volume of overflow in this period was 3,000,000 m^3 as detailed below:

June:	750,000 m^3	
July:	900,000 m^3	(of which 500,000 m^3 occurred on 2 July)
August:	150,000 m^3	
September:	1,200,000 m^3	

• **In the summer of 1994** the rainfall was significantly higher (280 mm during the same four months). The distribtuion of overflow—total 8,900,000 m³—was as follows:

June:	1,100,000 m³	
July:	3,800,000 m³	(of which 1,450,000 m³ occurred
August:	2,000,000 m³	on 18 and 19 July)
September:	2,000,000 m³	

Several samples of the overflow were analysed. The programme of measurements for one year for the Bièvre-Bas basin yielded the following values of concentrations of various compounds (averaged over Bièvre Watt, Blanqui-Gare, Buffon weirs):

SS:	200 mg/l
BOD:	40 mg/l
COD:	160 mg/l

It was also found that the concentration of SS increased significantly from upstream to downstream, undoubtedly due to resuspension of deposited matter in the network.

Earlier measurements made in 1979 by LROP on the Proudhon weir yielded significantly different values of the concentration of COD and BOD:

SS:	245 mg/l
BOD:	80 mg/l
COD:	275 mg/l

SAP also installed a station at Pont de l'Alma for observation of the quality of the Seine. It continuously measures six parameters including temperature and concentration of dissolved oxygen. A second station is planned farther upstream at Pont d'Austerlitz.

2. PROGRAMME OF MODERNISATION OF STORM WEIRS

2.1 PLAN OF MODERNISATION OF PARISIAN SEWERAGE SYSTEM

The situation described above is to be improved soon. In fact, SAP launched a plan of modernisation of the network in 1991; the main objectives are:
— reinforcement of the safety of the installations and equipment;
— total suppression of overflow during the dry season, even in the case of diversion works in the network, and reduction of overflow in the rainy season;
— rehabilitation of degraded works;
— modernisation of the methods of operation, in particular cleaning.

The reduction of overflow, combined with an increase in storage capacity, should help in preserving the quality of the Seine and preventing henceforth shock effects and fish mortality, continually observed during unusual storms. The managers of sewerage networks of districts adjoining Paris have also set similar goals under a directive administered by the Interdistrict Syndicate for the Sewerage System of the Parisian Agglomeration.

Three actions initiated under this plan pertain to storm weirs:
— installation of a network of permanent measurements,
— regulation of overflow bays,
— modernisation of closure devices of weirs in the Seine.
These three operations are detailed below.

2.2 NETWORK OF PERMANENT MEASUREMENTS

In 1994, SAP initiated a programme for equipping major works with a network of permanent measurements. The work will continue in 1995 and about 80 stations are planned which will be expeditiously equipped (limnimeters, flowmeters and rain gauges). It will then become possible to permanently know the hydraulic state of the network and to rapidly calculate the flow rate and volume of overflow.

Other stations will then be installed to achieve full-time monitoring of overflow, thereby meeting the conditions of the water law (deadline 3 January 1997). The results of these measurements will be communicated to a control centre and telesurveillance of the network studied.

2.3 REGULATION OF OVERFLOW BAYS

As mentioned above, weirs with fixed thresholds are activated as soon as the rainfall attains a certain mean value.[1] SAP therefore extended the plan to include the installation of automatic adjustable valves which will operate according to the·following principle:
— during the dry period the valve is adjusted to such a level that overflow does not take place;
— during light or medium rainfall the valve is raised as and when the water level in the sewer rises and overflow is thus prevented;
— if the rainfall is prolonged or becomes heavier and the water level rises to a limit beyond which there is a risk of overloading, the valve maintains its position and the overflow starts;
— should the water level continue to rise in spite of the overflow, the valve is lowered to increase the rate of overflow and the level maintained below the fixed limit.

---

[1]. An overflow taking place on two bays is shown in coloured plate no. 16.

Operation of the valve is controlled by the level gauges installed in the sewer and the weir.

About 20 storm weirs are planned to be equipped with this type of valve between 1994 and 1996. They will be automated at the first instance according to local requirements. Information pertaining to their state and the amount of eventual overflow will be transmitted to the network control station. Only at a later stage will a co-ordinated management of works be implemented and the valves so controlled that they can fulfil additional requirements in real time.

These valves will enable suppression of overflow due to light or medium rainfall and significantly reduce the volume discharged over the year. They will be less effective during strong storms because the network is then highly saturated and prevention of overflow without risking inundation is difficult. The programme of modernisation includes the construction of works of storage of rainstorm run-off to overcome this deficiency. The role of valves will not be insignificant, however; they will make it possible to partially divert the discharge towards the weirs connected with these works and thus optimise the storage capacity.

The works of storage will retain the overflow throughout the duration of rain and release it in the network leading to the treatment works at the end of the rainfall. There are two types of storage works: underground storage tunnels and storage basins.

Studies are in progress to install underground storage devices. Given the large volume required and clutter of the underground area of Paris, these studies are likely to be rather involved.

Hydraulic studies and measurements show that the largest overflow occurs in the eastern part of the capital; the concerned weirs will be given priority. Thus construction of the first work has been started on the Proudhon weir situated on the right bank, on the Pont de Tolbiac. This diversion basin for the weir has a volume of 18,000 m³; it will be installed under the planned Parc de Bercy.

2.4 MODERNISATION OF CLOSURE DEVICES OF STORM WEIRS IN THE SEINE

Protection of the network against floods of the Seine is presently secured by two methods:

— either by closing the mouth of the weir into the Seine by an adjustable barrage; this barrage is raised as and when the water level rises so that the possibility of overflow is not altogether precluded; this requires repeated and expensive operations during the flood period;

— or by closing the weir, as soon as the flood starts, with a watertight gate; in this case the possibility of overflow is nil.

The programme of modernisation includes replacement of these systems of closure with adjustable valves which are raised as and when the water rises. They will be operated according to the level of the river as measured by a gauge and all information pertaining to their state will be transmitted to the network control station. The station will also be able to manoeuvre the valve by remote control.

In some cases the same valve will be able to operate the closure mechanism in the Seine and regulate overflow. All weirs are to be equipped in this manner; the work will hopefully be completed in 1996.

The three 'maps' appended illustrate the modalities of hydraulic development and operation of 'Alma Site'.

Alma Site in Sewerage Network of Paris

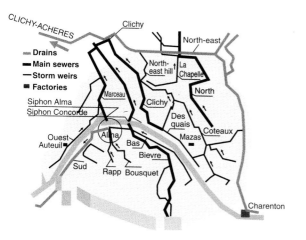

The waste water and rainstorm run-off flow naturally through the network of main sewers and drains towards the treatment works at Achères. This gravity network, designed at the end of the nineteenth century, can be quickly and easily modified to change the course of flow and natural operation.

The storm weirs are safety valves allowing operation of occasional diversion to the river in case of a violent storm or saturation of the network. The storm weirs protect the city against inundation.

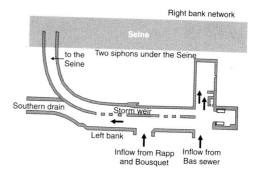

Alma Site receives part of the effluents of the left bank and divides it into two portions. One portion is collected on the right bank by passing under the Seine through two siphons, while the other rejoins the Southern drain.

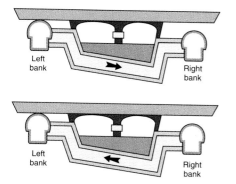

For correct operation of the network and the Alma siphon, the water level of the left bank should be higher than that of the right bank.

In the contrary case, the direction of flow would be reversed; the effluents of the right bank would be diverted from their natural course and carried towards the left bank.

The effluents can saturate the left bank network and thus cause overflow to the river by the storm weir.

193

Hydraulic Plan of Alma Site

Two objectives: Control of flow and protection of the Seine

Installation of automated equipment on Alma Site makes possible instant adaptation of the flux and maintenance of a constant water depth to ensure optimal utilisation of the works. A system of baffles for flotsam (bottles, plastics, polyesters, wood etc.) has been placed on the storm weir to prevent its entry into the river in case of overflow.

Operational Devices

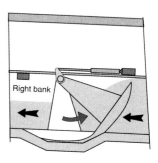

Drain Valve

This hydraulic valve regulates the depth of the water body at Alma

Right bank

Storm weir

Weir valve

This valve limits the overflow into the Seine by allowing maximum storage of effluents in the network in case of storm

Guillotine valves

These vertical valves close the two siphons to prevent passage of water of the right bank to the left bank in case of reversal of flow

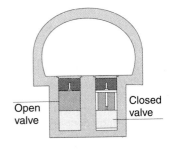

Open valve

Closed valve

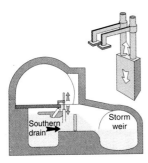

Southern drain

Storm weir

Baffle walls

Adjustable baffle walls are positioned vertically depending on water depth. They block all flotsam irrespective of level in the network and prevent its discharge into the river by the storm weir.

Right bank network

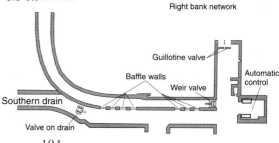

Guillotine valve

Baffle walls

Weir valve

Automatic control

Southern drain

Installation of equipment

Valve on drain

194

Hydraulic Operation of Alma Site

Automated control

The gauges measure the water level and transmit the information to a computer which pilots instantly and continuously the various valves of the Site.

In future, these automatic command devices will be integrated into a general management system for the entire networks and telecommanded from a central station.

Examples of scenarios

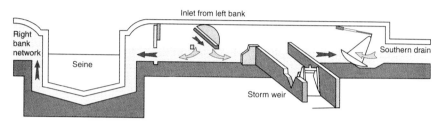

In normal operation the effluents are divided into two parts, one towards the drain and the other towards the siphons, by regulating the water level on Alma Site by means of the valve on the drain. The position of the storm weir valve precludes overflow into the river.

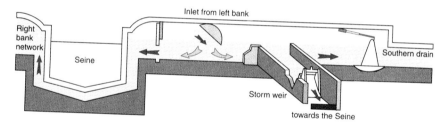

In case of a heavy storm, the storm weir valve plays the role of safety valve. Its position is adjusted to each situation, enabling prevention of saturation of the network by keeping the overflow minimal.

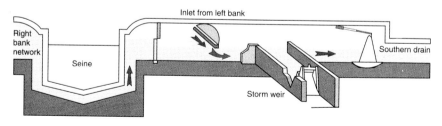

When the level of effluents of the right bank entrains a reversal of flow direction in the siphons, the guillotine valves close and prevent passage of effluents to the left bank. The position of the valves enables transport of the effluents of the left bank in toto to the drain. The weir valve always protects the Seine. In a heavy storm it would regulate the surplus to prevent saturation of the network.

195

Measurements on the
Storm Weirs of Parisian
B. Béthouart (City of Paris)

1. PREAMBLE: CHARACTERISTICS OF PARISIAN WEIRS

In Paris, a 'weir' is defined as a work which collects the overflow of combined sewers, the excess flow upstream or even backflow of pumping stations and the manholes situated along its stretch.

It can thus receive flow from one or several sources.

The overflow generally comes from a series of bays (1 to 13 depending on the case) whose thresholds are formed by wooden beams. The overflow is generally lateral with respect to the flow of the sewer. Some bays have been replaced by sector valves under a plan of automation.

Also, the weirs include a closure system in the Seine constituted of various devices (sector valves, gate with a float, barrages with metallic beams etc.) downstream of the weir, which prevent intrusion of Seine water into the sewerage system when the river is in flood.

In general, in view of the distance between weirs, the closure system in the Seine is distinct from the inlets to the weirs.

2. APPROACH USED IN A PROGRAMME OF MEASUREMENTS INITIATED IN 1993

2.1 Determination of theoretical capacity of weirs

The Section of Sewerage of Paris used models based on the results of various studies. They enabled determination of the volume of discharge for a decennial rainfall for each weir or for a combination of weirs. This volume is treated as a criterion for categorisation of relative weir size.

2.2 SELECTION OF WEIRS TO BE EQUIPPED

The goal was to measure 80% of the volume of overflow by equipping a minimum number of weirs; the choice was based on relative size and feasibility of measurements.

By and large, we conducted measurements on all weirs whose volume was larger than or equal to 3% of the total volume, with the following three exceptions:

— the weir of the eastern periphery which is not under the management of SAP and mainly drains rainstorm run-off (not combined);

— the weirs at Marine and Mazas whose claimed discharge is tied up with the weirs at Diderot, Bercy and Traversière because the management of these five weirs is interdependent; thus all five weirs would have to be equipped though on average the overflow per weir is less than 2%.

In addition, we equipped the Wilson weir and the common section of the Chatillon Bas Meudon and Renan Seine weirs because they are situated at the limit of the models and there was considerable uncertainty regarding their functionality.

Lastly, we equipped the Javel Leblanc weir because we apprehended that a division between this weir and the Javel Convention weir would lead to results differing from those of modelling.

2.3 CONSTRAINTS TAKEN INTO ACCOUNT

2.3.1 Measurement of inflow from combined or partially combined sewers

In some cases the manholes are connected with the weir. We did not plan to conduct measurement on the most downstream manhole. On the other hand, where there were several large combined inflows into a weir, we measured them except in the case of the Proudhon weir when a secondary sewer draining into a very small catchment area (1 to 10 ha) was not taken into account.

2.3.2 Choice between conducting measurement in the weir and in the sewer at the level of bays of overflow

Bays of overflow are unique and formulas are available in the literature for calculating the rate of flow passing through them in terms of depth upstream of the bay (in the sewer) and eventually downstream. However, these formulas are based on hypotheses which are seldom or rarely valid in the case of Parisian weirs.

In fact, on the one hand the flow in the sewer is more or less parallel to the overflow bays (the sewer can change direction on the weir or the discharge of a secondary sewer can face the bays) and on the other, the sills can function as immersed weirs.

Lastly, regulation of these sills is carried out by wooden beams whose total number and height can be varied according to the user requirement or that of the yard or even left as is depending on intensity of flow. To obtain the flow rate by measuring the depth in the sewer requires reliable knowledge of the position of these beams. Hence measurement is preferably conducted in the weir.

Also, if a weir includes several large inlets, this method can be used to reduce the number of points at which measurements have to be conducted.

2.3.3 Position of last overflow with respect to measurement site

If measurement is conducted in the weir, the site of measurement must be at a sufficiently long distance from the last overflow because the turbulence generated causes fluctuation of the wet sections and it is not possible to determine flow rate by measuring velocity.

2.3.4 Position of last overflow with respect to the river

If possible, measurement should be conducted in a region uninfluenced by the river, which is imperative if the flow rate is to be ascertained by measuring water depth.

The influence of the river necessitates measurement of the velocity but this measurement does not always suffice. If the section wetted solely due to the influence of the river is large, detection of overflow becomes problematic. In fact, if a velocity probe based on the Doppler effect is used, the minimum velocity measured being about 20 cm/s, the minimum flow rate calculated for a totally immersed D 1000 (or partially filled D 1400 or even a D3000 with 50 cm of water) will be about 160 l/s or 13,000 m³/day.

When the actual overflow is smaller than this value, the probe gives faulty readings and a manual and systematic validation of this type of measurement becomes necessary.

This constraint was taken into account only for a river at low water level and the programme of measurements was initially confined to the summer of 1993.

2.3.5 Case of river floods

The constraints mentioned above change with the water level of the river. Thus the unaffected sections may represent this level while the wetted sections may increase it.

In fact, so long as the weir is not completely closed, it can function; this has been confirmed by the management of Parisian weirs. This may be acceptable, however, as long as the duration of invalidated measurements is shorter than a threshold, yet to be defined in terms of the accuracy required for these measurements.

Should measurements of weirs be undertaken when the river is in flood? If so, up to what intensity of flood? It is known that some weirs never completely close while others close partially at levels lower than the one which generally invites follow-up action.

As for the programme of measurements conducted at Paris, this constraint invalidated many measurements made in the winter of 1993/94.

3. SOLUTIONS ADOPTED

3.1 LAW OF UNSUBMERGED WEIRS

This law or formula was adapted in the algorithm of compuation to determine flow rate from the water depth measured upstream of the threshold(s) where several unsubmerged weirs are situated.

In the Javel Leblanc case the threshold level was the same as the weir level. Except for Seine floods, unsubmerged weirs may be expected to function.

3.2 STRICKLER'S LAW

This formula enables computation of the flow rate from the measured value of depth in that case wherein the stream line parallels the floor (uniform flow). Calibration of each site enables integration of friction and the actual slope (for the conditions of calibration).

3.3 MEASUREMENT OF ENERGY LOSS

This method corresponds to the particular case of St Michel weir whose largest section on the quasi-totality of the weir length is terminated by 3 D 400 mm. It enables computation of the flow rate from variation in the measured value of depth upstream of the conduits. It is based on the hypothesis that the depth downstream of the conduits remains constant during a storm event and is equal to the depth upstream of the conduits before the event.

However, this method is not applicable in the case of small flow rates because the variations of the downstream water body (lapping in the Seine) make measurement of the head loss difficult when less than 30 cm.

In practice, a second measurement of depth enables detection of the start and end of overflow and determination by interpolation using a curve H^b (between 0 and a calibration flow rate) of the small flow rates.

3.4 DIRECT MEASUREMENT

This method consists of determining the wetted section from the measured values of depth and then computing the flow rate using the law $Q = SV$, where V is the flow velocity.

In the past, the method of the Doppler effect was used but today velocity is computed by measuring the time of transit along a path.

3.5 OTHER APPROACHES

In some cases measurement of depth was repeated to increase reliability.

In the case of a weir with an inlet threshold and influence of the Seine, a gauge for measurement of depth enables detection of overflow and validation of measurement of velocity in the section.

Details of the measurement programme executed at 14 sites are given in Fig. 1.

4. PROBLEMS ENCOUNTERED

4.1 AVAILABILITY OF MEASUREMENTS OF DEPTH AND VELOCITY

4.1.1 Measurement of velocity by Doppler effect (ADS)

Velocity was measured directly on the weirs for which the point of measurement could be or was influenced by the level of the Seine even in summer (DO Buffon, DO Blanqui, DO Trois Baies, DO Javel Convention).

Numerous measurements were rejected and computation of volume sometimes necessitated installation of a depth gauge in the sewer (DO Trois Baies) to confirm the actuality of an overflow.

Computation of volume and duration of overflow was done using ADS. The controls excercised from the selected measurements obtained from ADS always gave different values.

Weir	Method	Quantities measured	Instruments used
Alma, Right Bank	Threshold law	Depth upstream of a threshold	US aerial gauge 'Aqualyse'
Alma, Left Bank	Threshold law	Depth upstream of a threshold	Pressure gauge 'CR2M DRUCK'
Bièvre	Strickler's law	Depth on weir	Pressure gauge 'CR2M DRUCK'
Blanqui	Direct measurement	Depth and velocity on weir	Pressure gauge and US aerial gauge
Buffon	Direct measurement	Depth and velocity on weir	Pressure gauge and US aerial gauge
Bourgogne	Strickler's law	Depth on weir	Pressure gauge CR2M DRUCK and US aerial gauge MILTRONIC
Chatillon Bas Meudon and Renan Seine	Strickler's law	Depth on weir on common section	Pressure gauge CR2M DRUCK and US aerial gauge MILTRONIC
Javel Convention	Direct measurement	Depth and velocity on weir	Pressure gauge and US aerial gauge
Javel Leblanc	Threshold law	Depth upstream of a threshold	Pressure gauge CR2M DRUCK
Proudhon	Strickler's law	Depth on weir	Pressure gauge CR2M DRUCK and US aerial gauge MILTRONIC
St Michel	Head loss and $Q = aH^b$ law	Depth at two points on weir	Two pressure gauges CR2M DRUCK
Trois Baies	Direct measurement	Depth and velocity on weir	Pressure gauge and US aerial gauge
Vincennes Charenton	2 Strickler laws	Depth on weir	Pressure gauge CR2M DRUCK and US aerial gauge MILTRONIC
Wilson	Strickler's law	Depth on weir	Pressure gauge CR2M DRUCK and US aerial gauge MILTRONIC

Fig. 1.

4.1.2 Determination of velocity by measuring transit time (CR2M)

The Trois Baies, Javel Leblanc and Javel Convention weirs were later equipped with this type of instrumentation. To date, only weir closure has been measured.

4.2 Reliability and accuracy of depth measurements

The accuracy of measurement of depth intervenes at two levels:
— measurement of the wetted section irrespective of the method of measurement,
— determination of velocity for the depth-flow rate relationships.
Hence the importance of correct measurement of depth and in particular a correct relative accuracy in the case of small depths.

As a matter of fact, when the depth is small the sectional area generally varies rapidly (more or less circular shape). Thus for D = 3000, error in the area is more than 20% when the depth of flow is less than 18 cm (respectively 86 cm) and the error in the measurement of area is 2 cm (respectively 10 cm).

This phenomenon is accounted for in the computation of volumes by the frequency of small depths with respect to the set of measurements.

4.3 Methods of measurement of flow rate
not suitable for certain sites

4.3.1 DO Bourgogne

The choice (computation of flow rate by a Strickler law) appeared to be unsuitable for two reasons:
— during calibration the depths and flow rates obtained were not coherent with the theoretical values of K_s and the local slopes: the values obtained were too large and we preferred to retain the theoretical values;
— during the programme of measurements the obtained values of flow rate and volume were often invalid because the duration of the overflow (several days while the other weirs functioned only for a few hours) and its volume were not coherent with those of other works for which measurements were conducted.

4.3.2 DO Vincennes Charenton

This is the largest Parisian weir and the only one with permanent overflow; it is therefore imperative that the rate of discharge be measured. It was found that the Strickler laws were limited during Seine floods when this work was put into operation. Measurement of velocity would have yielded better knowledge of its functional capability.

4.3.3 DO Alma, left bank

Modifications of the weir, consisting of the addition of an adjustable threshold level and mobile siphon partitions, necessitated modification of the computation of flow rate, earlier calculated using the frontal threshold law with a mean coefficient of overflow for each bay.

As the outlet sewer itself is placed much lower than the water level of the Seine, measurement of velocity did not suffice (overflow less than 600 l/s not detectable). A brace of solutions was necessary, in which measurement of velocity proved helpful in calibrating the functionality of bays during large flow rates.

4.3.4 DO Javel Leblanc

The method used was that of a threshold with thin walls in the weir. This method has a drawback, namely, it is valid only for a relatively small section of the banks of the Seine. Further, it cannot be used for programmes of short duration.

5. INTEGRATION OF DEVELOPMENT OF WEIRS

5.1 SECTOR VALVES ON OVERFLOW BAYS

Generally management of the sector valve stipulates measurement of the rate of overflow (passing over the valve). On the other hand, the valves on overflow bays can open completely and therefore the weir operates in submerged conditions, which necessitates measurements both upstream and downstream of the valve and knowledge of the valve's position. All this information should be recorded in a synchronised manner to enable computation of the flow rate. When the valve is completely lowered, the law is not valid.

In practice, calibration of the rainy period should be planned, which implies a programme of direct measurements (depth and velocity) of the rate of overflow whose duration depends on the sensitivity (frequency of functioning) of the weir.

5.2 DEVICES AGAINST FLOTSAM

The fixed siphon barriers also enable computation of the flow rate but not the adjustable barriers (whose position is too random). In this case, also, it is necessary to plan calibration in the rainy season.

6. CONCLUSIONS: SOLUTIONS SUGGESTED FOR THE NEXT SITES OF MEASUREMENTS ON WEIRS

6.1 ESSAY MEASUREMENT OF DEPTH ONLY BUT TAKE TWO MEASUREMENTS AT EVERY SECTION (4 GAUGES IF MEASUREMNTS ARE CONDUCTED AT TWO SECTIONS TO COMPUTE THE FLOW RATE)

Aside from giving a better accuracy, especially at a small flow rate, measurements become automatically valid.

6.2 MODEL THE RAINY PERIOD

Velocity is measured on a section of the weir to determine the rate of overflow. This enables:
— accuracy of direct measurements on the entire range of values obtained;
— reliability, availability and expeditious validation of measurements of depth once the system is calibrated.

This principle assumes that the storm events of short return periods would generate a significant flow rate on a representative portion of the range of potential measurements. Otherwise, this programme of calibration should be replaced by one of permanent measurement of depth and velocity (permanent calibration).

6.3 CONDUCT MEASUREMENT OF DEPTH IN THE SEWER FOR DETECTION OF OVERFLOW IN THE CASE OF A SINGLE INLET

This principle is helpful in that either only depth need be measured to compute a flow rate or measurements of velocity validated.

Measurement of Water Quality in a Sewerage Network

A. Rabier[1], M. Van der Boght[2] and M. Wollast[2]

INTRODUCTION

Since the end of April and partial launching of the programme 'SEINE-PROPRE', part of the effluents of the sewerage networks of Val-de-Marne has been diverted to the treatment works at Valenton (section 1b). However, the equipment installed by the Directorate of Services of Water and Sewerage (DSEA) of Val-de-Marne instantly redirects the whole or part of these effluents towards the collection and treatment works of Achères (STEP) and thus the situation reverts to its earlier state.

This equipment forms part of the VALERIE system of management responsible for regulating the effluents to the Valenton plant and thus enables compliance with the agreement between Val-de-Marne and SIAAP (Interdepartmental syndicate for the sewerage of the Parisian agglomeration). It also helps to meet the needs of management of SIAAP and users of the Valenton treatment works.

This hydraulic diversion of 130,000 to 150,000 m³/day of effluents towards the STEP of Valenton is only one aspect of the agreement. The second aspect relates to the qualitative regulation of these effluents and necessitates acquisition of knowledge of not only flow rate, but also of the quality of the effluents.

Knowledge of the flux of pollutants was obtained in the dry periods of 1988 to 1992 by 12 meticulous sampling programmes from the various sewers concerned. The arduous and long operations conducted on site and in laboratories do not provide knowledge in real time, however, of the quality of the effluents. Thus since 1991 the DSEA has been carrying out measurement of the quality of effluents in the network, the requirements being a reliable value and the process of obtaining it automatic and continuous. Since the end of 1992 a measurement station has been operative on work XIV at Creteil.

---

[1] DSEA Val de Marne
[2] Aquarius Society

This paper includes a description of the instrument used and the information obtainable by it as well as the various stages of validation of the series of measurements made. The problems of management of the instrument are tackled and its planned use in the basin for storage and treatment of water at Vitry-sur-Seine is discussed.

1. THE ADES INSTRUMENT

This instrument (shown in Fig. 1), marketed by the Aquarius Society, was designed in collaboration with the Open University of Brussels. It has been extensively tested over the years in the domain of continuous measurements on rivers as well as sewerage networks. The analyser ADES is a modified version of the instrument installed on the main sewers of Brussels between 1983 and 1989.

1.1 GENERAL DESCRIPTION OF THE INSTRUMENT

The instrument comprises the following subassemblies (Figs. 2 and 3):
— **A reservoir of measurements,** a cylindrical chamber equipped with various probes and gauges needed for determination of temperature, conductivity and turbidity. On the lower part of the reservoir a sampling tube, for removing waste water for analysis, is connected to a valve with a cylindrical body.
— **An electronic compartment,** comprising the power connection, electronic circuits for measurements as well as electronic circuits of acquisition, storage in memory, command and communication with the world outside.
— **A pneumatic compartment,** a set of valves and instruments for regulation (regulators, manometers etc.).
— **A set of pumps:** vacuum pump, compressor and an interface module.
— **A reservoir of compressed air.**
— **Two reservoirs,** one for the **control fluid** and the other for the **cleaning fluid.**
— **Three valves for inflow** of the liquids entering the reservoir of measurement: one valve for inflow of control fluid, one for city water and the third for the cleaning fluid.
— **A junction box** containing the set of units for the electrical connections of the analyser (sector, digital inlet and outlet, transmission line) as well as the devices of electrical protection (circuit breaker and fuses).

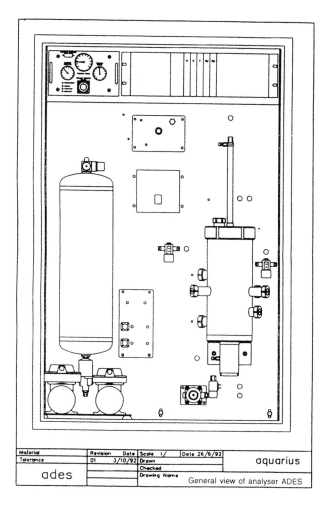

Fig. 1

— **A level gauge** for measuring water depth.

All these parts are readily accessible from the front of the analyser.

1.2 OPERATION OF ADES

The operation of ADES is relatively simple; it can be described in terms of a sequence of activities or 'jobs'. Each job is itself made up of a succession of 'tasks'. In the normal mode of operation, three jobs are executed by ADES: measurement, washing, sampling at intervals. Other jobs can be performed at the demand of the operator and variations from the mode of

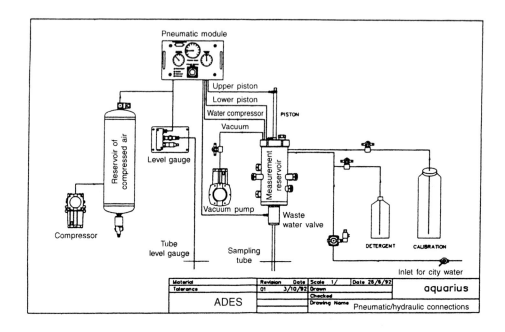

Fig. 2

normal operation can be executed when the analyser detects the appearance of particular situations.

We shall confine ourselves to a simplified description of methods used and various tasks executed by ADES in the normal mode of operation.

1.2.1 Sampling and measurements

Although one generally speaks of continuous measurement of the quality of effluents, ADES operates according to a discontinuous mode of sampling, with physicochemical analyses carried out in a measurement reservoir equipped with various necessary gauges.

It can operate in normal mode (long cycle, for example a measurement lasting half an hour) or in fast mode (short cycle, say a measurement every five minutes). The changeover from normal to fast mode can be engineered either on command or automatically when the rate of rise of water level exceeds a preassigned value.

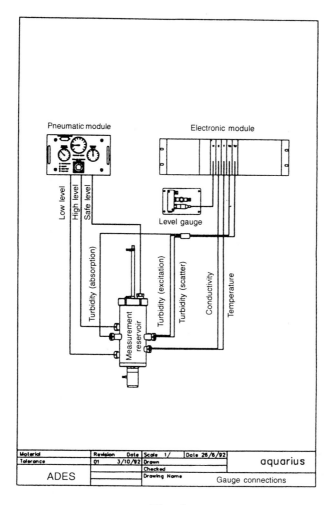

Fig. 3

The sample of water (1.5 l) to be analysed is obtained by suction from near the bottom of the reservoir after opening the valve.

In the standard version of ADES a sample is drawn from a depth of about 7 m which is the distance between the level when the reservoir is full and the free surface of the medium to be analysed; thus only the tip of the suction tube needs to be immersed.

To guard against blockage by various types of solids present in the waste water, the sampling tube is sufficiently wide (internal diameter 17 mm) and the main valve situated at the base of the reservoir is made of elastic 'fur', the diameter of which in an open position is equal to that of the sampling tube. Furthermore, ADES employs an automatic procedure for evacuation and purging under pressure.

It is futile to place a strainer (or any other sieving or filtration device) at the submerged end of the sampling tube. A common source of clogging is thus eliminated.

Once the reservoir has been filled, measurement of physicochemical parameters (temperature, conductivity and turbidity) is conducted after a stabilisation period of a few seconds. Various gauges are deployed at two levels along the lateral wall of the reservoir: at the lower level the temperature and conductivity gauges are mounted on the right side of the reservoir and at the upper level the device for measurement of turbidity, light source and cell for measuring scattered light. The cell for measuring transmitted light is placed on the left side. The plane of measurement of turbidity is at a distance of 5 cm from the free water surface at the end of the filling cycle.

After recording the results of measurements conducted on waste water, ADES initiates a phase of static decantation during which the value of turbidity is measured at regular intervals. This phase of measurement enables characterisation of properties of decantation of suspended matter present in the sampled water and quantification of the rapidly decantable fraction.

Upon completion of the cycle of decantation measurements, evacuation of the reservoir takes place by reversing the flow in the sampling tube under the action of compressed air. This purging under pressure precludes contamination of successive samples by residue from earlier samples. Clogging of the sampling tube is likewise prevented.

To ensure that the sample is as representative as possible, two precautions are exercised: on the one hand, the flow rate of the vacuum pump, operated to fill the reservoir, enables obtaining a velocity of water circulation in the sampling tube higher than or equal to 1 m/s. A velocity of this magnitude guarantees representative sampling with respect to particulate matter and excludes any decantation or sedimentation. On the other hand, each phase of measurement is preceded by a first-cycle rinsing with waste water (filling-evacuation-purging) to eliminate the influence of the liquid priorly present in the reservoir.

1.2.2 Calibration and cleaning

ADES deploys a system of valves that enable filling the reservoir, either with city water or with a control fluid. In the normal mode of operation, filling with city water at the end of each cycle of measurements eliminates eventual sedimentation on the walls and gauges. After evacuation of the rinse water, the reservoir is again filled and this terminates the normal cycle of measurements.

The gauges are maintained under clean water conditions during the waiting period between two cycles of measurements. To initiate the next cycle of measurements this clean water is used for preliminary calibration of the turbidimeter by bringing the signal to zero turbidity (measure of 'zero turbidity').

A more complete cycle of cleansing should be executed at regular intervals to prevent gradual fouling of the reservoir walls and gauges. ADES does this through chemical and mechanical means.

When a washing phase is initiated, a small volume of cleaning fluid is introduced in the reservoir, which is subsequently filled with city water; walls and gauges are then scrubbed with a brush. This brush, usually kept in the upper part of the reservoir above the maximum level of filling, executes a series of vertical movements under the action of the pneumatic cylinder mounted on the lid.

The washing cycle concludes with four successive rinsings with city water in order to eliminate all residues of the cleansing substance.

Lastly, ADES can accomplish a cycle of calibration by creating a vacuum in the reservoir and sucking a sample of the control fluid. The fluid is a standard solution of potassium chloride (KCl) enabling monitoring of the conductivity gauge and verification of the initial setting (zero turbidity) of the turbidity gauge. To preclude contamination of the control fluid, it is discharged after use.

1.3 INFORMATION OBTAINED FROM ADES

The system of ADES measurements provides two types of information:
- Direct measurements obtained from the specific gauges:
  — measurement of temperature with an accurate thermistor;
  — measurement of conductivity by means of a cell with four platinum electrodes;
  — measurement of optical properties (absorption and scatter) of suspended matter. The light intensity resulting from the interaction of incident light with the medium (sample) is measured by two detectors forming angles 0° (absorption) and 135° (scatter) normal to the emitter. The principal wavelength of excitation is 637 nm and is thus in the red part of the spectrum. The angles of observation of the sample by the detector cell are limited to about 8° by collimation (Fig. 4). The measurement of optical properties is conducted on raw water and water in the course of decantation;
  — measurement of water depth in the main sewer by means of a pressure gauge linked to a pneumatic channel whose end is submerged in the minimum level of flow ('bubble to bubble' system of measurement).

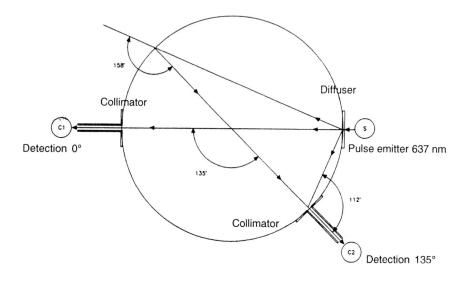

Fig. 4

- Derived information, obtained by analysis of direct measurements:
  — suspended matter (total or decantable SS) from measurement of optical properties;
  — organic matter (total, particulate or dissolved) expressed in terms of COD by combination of measurement on raw and decanted water (optical measurements and conductivity).

Compared to direct measurements, derived information constitutes estimates based on a certain number of hypotheses pertaining to the physical properties of suspended matter and the behaviour of flow of waste water of dry and rainy seasons.

To completely validate the set of information (direct and derived) provided by ADES, the following three steps are necessary.

1. Initial calibration (in the laboratory) of the gauges used.
2. Determination of numerical coefficients involved in the various relations used for estimating the derived parameters (SS and COD) from direct measurements.
3. Validation of the data set by in-situ intercalibration

2. INITIAL CALIBRATION AND INTERCALIBRATION

2.1 INITIAL CALIBRATION OF GAUGES

2.1.1 Temperature

Laboratory calibration of the temperature gauge is carried out in two steps:
— verification of the resistance-temperature curve provided by the manu-
facturer by means of a multimeter and an accurate thermometer;
— plotting the response curve of the electronic meter (signal obtained as
a function of the resistance).

These two operations enable determination of the values of the following
coefficients:
— coefficients for computation from raw data (resistance R in terms of
value S of the signal)

$$R = a_0 + a_1 S + a_2 S^2 + a_3 S^3 + a_4 S^4$$

— coefficients for computation from physical data (temperature T in terms
of resistance R)

$$T = b_0 + b_1 R + b_2 R^2 + b_3 R^3 + b_4 R^4$$

These procedures are verified by direct comparison (between 0 and 30°C) of
the temperature obtained from ADES and that shown by a certified accurate
thermometer.

2.1.2 Conductivity

Laboratory calibration of the conductivity gauge is carried out in two steps:
— plotting the response curve of the electronic meter (signal obtained
versus conductivity) by means of an accurate resistance box;
— determination of the constant of the cell of the conductivity meter by
means of standard solutions (KCl).

These two steps enable determination of the values of the following coefficients:
— coefficients for computation from the raw data (conductance in terms
of the signal S obtained)

$$\Lambda = a_0 + a_1 S$$

— coefficients for computation from the physical data (conductivity in
terms of conductance)

$$\text{conductivity} = b_0 + b_1 \Lambda$$

215

Measurement of conductivity is related to the reference temperature by application of a correction in temperature by the following law:

$$C (T_{ref}) = C (T)/ [1 + 0.0194 (T_{ref} - T)]$$

2.1.3 Optical properties

The response of optical gauges is calibrated in the laboratory by insertion of grey filters of known optical density (OD between 1 and 4). For such calibration two gauges are placed, in turn, in the axis of the light source.

This operation enables determination of the values of coefficients required for computation from raw data (absorbance and coefficient of scatter in terms of the signal received).

2.2 ESTABLISHMENT OF RELATIONS BETWEEN DIRECT MEASUREMENTS AND DERIVED PARAMETERS

Two approaches are used for establishing these relations.

2.2.1 Estimation of SS

The relation between measured optical properties and concentration of SS is established from a model taking into account the geometry of the system of measurements and the optical characteristics of the light source and detectors. The relation obtained by modelling is subsequently validated on standard suspensions. We used kaolin suspensions (granulometry: 1 μ; sp. gravity 2.4) of a concentration between 0 and 3500 mg/l. The results of modelling and comparison with the values obtained by ADES are shown in Fig. 5.

It can be shown that the geometry of the measurement device enables determination of the coefficient of total scatter b due to particulate matter, for total absorbance of the suspension contained between 0 and 4 and this can be done without introducing error due to the coloration of the water—a common problem with experiments using only one of the two measurements. The relation between the coefficient of total diffraction determined by this method and the concentration of SS in the uncoloured water is linear between 0 and 3 g/l of SS (Fig. 6).

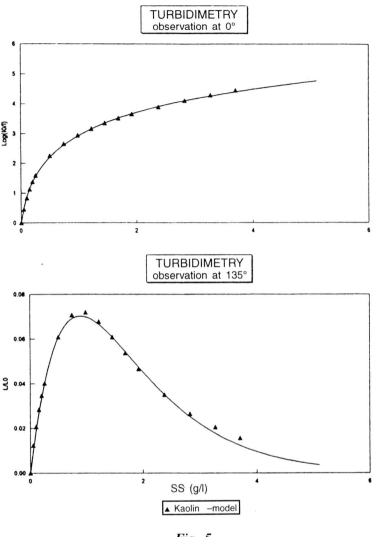

Fig. 5

2.2.2 Confirmation on waste water, speciation of SS and determination of COD

Intercalibrations on real waste water are subsequently carried out in order to:
— verify the general character of the absorbance-SS relation established according to the procedure described above;
— establish the relations that enable determination of mineral and organic fractions of SS and particulate COD (determination of dissolved COD will be discussed in Sec. 2.3).

217

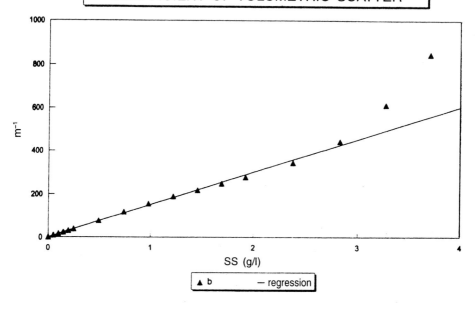

Fig. 6

This work of intercalibration was carried out during the programme of measurements on the principal drain of the city of Brussels (combined main sewer carrying predominantly domestic waste water with a mean flow rate of 1.2 m³/s, draining a basin of 2100 ha and a population of 350,000 inhabitants).

As part of the programme of measurements, synchronous samples were taken to feed, on the one hand, an ADES system (sample size: 1.5 l) and, on the other, a sampler of the MANNING S4040 type, ordered by ADES (sample size: 500 ml).

It is important to emphasise the fact that the automatic analyses carried out by ADES and the laboratory measurement did not use the same samples. Taking into account the short duration fluctuations in composition observed in the flow of waste water, the procedure followed may result in large discrepancies between the two sets of data—ADES and laboratory; this point is elaborated below.

The laboratory analytical techniques are as follows:

— Conductivity: measurement by means of a laboratory conductivity meter after calibration of the cell on the standard KCl solution.
— Suspended matter: by filtration through a fibreglass filter WHATMAN GF/C predried and preweighed (drying at 105°C and weighing by a

218

precision balance). The volume of filtered sample, varying with concentration of SS, is such that the increase in weight of the filter is of the order of 10 mg.

Remark: The samples are strained before filtration (sieve opening: 5 mm) to remove gross pieces (paper, fibres etc.) whose representativity is not certain because of smallness of the filtered volume (from 10 to 50 ml).

— Volatile matter in suspension: by calcination at 550°C and weighing the filter.

— COD: the total COD is measured on the sieved raw sample by the method of tube digestion followed by a spectrophotometric measurement of the Cr^{3+}. The same method is applied on the filtered water for determination of the dissolved COD.

The difference (total COD – dissolved COD) gives the value of the particulate COD.

An example of the results obtained under the programme of intercalibration on the drain at Brussels is given below. It may be noted that in this programme the ADES system currently in service at Creteil (work XIV) was used to obtain 'automatic' measurements under the framework of tests for commissioning the equipment on site before its delivery.

The two graphs (Figs. 7 and 8) pertaining to measurement of conductivity are useful in two ways. They demonstrate a good general concurrence between the two data sets but foremost they enable determination of a mean value of discrepancy between the measurements attributable to the different procedures followed (distinct measurements on synchronous samples).

In fact, it may be assumed that the accuracy of the measurement of conductivity is not affected by an interpretation of the raw signal. The principal source of error is related to the exact strength of standard solutions used for calibration; error from other sources (electronics, thermal compensation, dimensional stability of gauges, interpolation etc.) can be neglected in the first approximation.

In these conditions, the accuracy of the measurement of conductivity is of the order of 1%: a deviation larger than 10 S cm^{-1} between the ADES and laboratory measurements is a sign of a real difference in the composition of the two synchronous samples. Up to within the instrumental error, the dispersion of experimental points (Fig. 7) is thus essentially attributable to the difference in sample composition resulting from the sampling procedure. This dispersion may be characterised by the value of the coefficient of correlation r between the two data sets. *In the case of conductivity, the value of the coefficient of correlation between measurement on the Brussels site and the values obtained by the method of intercalibration, is 0.96.*

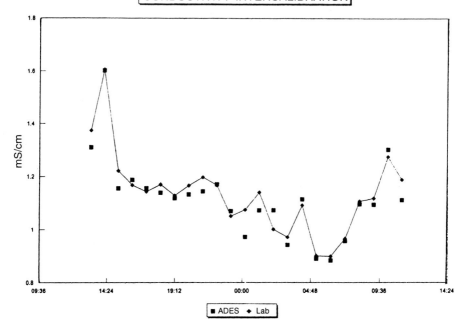

Fig. 7

Three graphs (Figs. 9 to 11) show the results of analysis of SS, particulate organic matter (POM) and particulate mineral matter (PMM). The data on SS provided by ADES (squares in Fig. 9) are the result of direct application of coefficients obtained from modelling and verified in the standard suspensions.

The speciation of SS in particulate and organic mineral matter is based on the hypothesis that the rapidly decantable solids are representatives of the gross fraction having a predominance of mineral matter, most of the POM being associated with the finer fraction possessing a smaller velocity of decantation. The POM obtained by ADES was thus calculated from the optical properties of the suspension after 100 seconds of static decantation by application of the same procedure of calculation used for total SS. The PMM can be determined by subtraction.

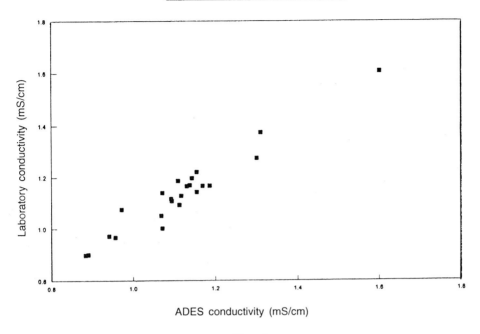

Fig. 8

Calculation of the coefficient of correlation for the sets of equivalent data yields the following values.

ADES SS/Lab SS: r = 0.96

ADES POM/Lab POM: r = 0.97

ADES PMM/Lab PMM: r = 0.78

For SS and POM these values, very close to those calculated for conductivity (r = 0.96), show that the dispersion of experimental points (Figs. 10 and 11) is not significantly larger than the dispersion characterising the measurement of conductivity. On the contrary, the largest dispersion observed for mineral SS most probably results from the fact that the hypothesis on which the calculation is based is not quite tenable.

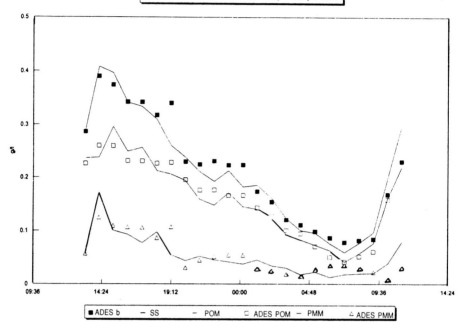

Fig. 9.

Thus the hypothesis need not be discarded altogether provided one knows that:
  - the difference in composition of synchronous samples constitutes the principal factor of observed discrepancies;

  - the procedure of calculation of SS from the optical properties can be generalised to cover the suspensions present in waste water with a predominant domestic component;

  - the matter decantable in 100 seconds correctly represents the mineral fraction of SS. (This decantation time corresponds to the complete removal of particles whose velocity of fall is greater than or equal to 1.8 m/h. For a sp. gravity of 2.5—characteristic of the mineral fraction—the diameter of these particles is larger than or equal to 0.025 mm.)

222

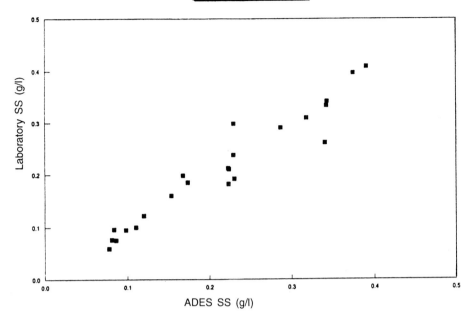

Fig. 10.

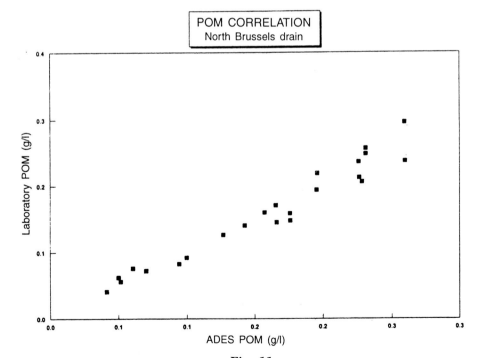

Fig. 11.

223

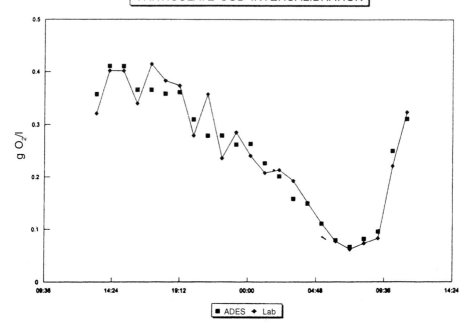

Fig. 12.

Evaluation of particulate COD is presented in Figs. 12 and 13. For this evaluation a linear relation between POM and particulate COD has to be assumed, implying that the COD equivalent of POM remains constant at all times. The assumed linear relation takes this form:

$$\text{particulate COD (mg } O_2/l) \ = \ 1.58 \text{ POM (mg/l)}$$

Calculation of the coefficient of correlation between the sets of data on POM as measured by ADES and that in the laboratory ($r = 0.97$) again indicates that the small discrepancy between them is attributable to the difference in composition of synchronous samples.

2.3 VALIDATION BY IN-SITU INTERCALIBRATION

This section discusses the programmes of intercalibration carried out between October 1992 and March 1995 for validating the information provided by the ADES measurement system installed at Creteil (work XIV). The main sewer carries about 27,000 m^3/day in the dry season with a mean concentration of COD and SS respectively equal to 610 and 276 mg/l.

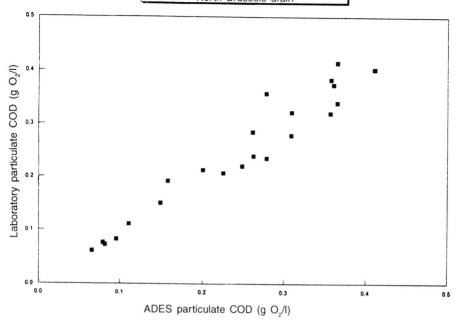

Fig. 13.

Table 1 gives the results obtained by several programmes of sampling and the date of execution of the programme.

Programme serial no.	Date	Time: 1st sample	Time: last sample	No. of samples	Laboratory
1	19-20/10/92	16:34	08:17	16	DSEA
2	03-04/11/92	18:10	01:20	8	DSEA
3	30/1/92 01/12/92	10:09	22:09	23	DSEA
4	12-13/12/93	16:00	21:45	5 (average)	DSEA
5	05/01/94	08:00	19:45	6 (average)	DSEA
6	11-12/01/94	16:00	03:45	9 (average)	DSEA
7	02/02/94	11:00	22:45	5 (average)	DSEA
8	30-31/03/94	13:10	11:10	23	ULB*
9	07-08/03/95	09:56	08:59	24	DSEA

*Laboratory of water treatment, Open University of Brussels.

Table 1

In programme nos. 4 to 7 measurement of the pollution parameters was conducted on samples averaged in terms of flow rate; corresponding results have not been included in this report.

The results of other programmes concern instantaneous samples and are thus directly comparable with synchronous measurements obtained by ADES. We have used both data sets (measurements by ADES and the laboratory) for programme nos. 1, 3, 8 and 9, representing a total of 86 synchronous samples. Programme no. 2 (8 samples) does not include the ADES results.

2.3.1 Sampling methodology

Except for programme no. 8, the samples for laboratory analysis were drawn under the supervision of DSEA/EA.08 using an ISCO sampler (by peristaltic pump). For programme no. 8, Aquarius used the MANNING sampler, model S4040 (vacuumetric operation). The sampling tube is permanently installed next to the water inlet of the ADES system.

Thus laboratory analyses and measurements by ADES are not conducted on the same samples, which may account for the divergence between the data sets. To minimise such differences, the sampler is controlled by the ADES system. The time-lag between two samples is approximately 30 seconds.

2.3.2 Laboratory analysis

For all sets of samples treated, the following information is obtained:
— conductivity, measured at the laboratory temperature, is converted to a value at the reference temperature 25°C;
— suspended matter, by centrifugation, drying at 105°C and weighing (for set no. 8, sieving and filtration on preweighed filter GF/C, drying at 105°C and weighing);
— total COD (on undecanted raw water, except for set no. 8 for which strained raw water was used) and dissolved COD (on filtered water).

It may be noted that the two methods of measurement of SS (centrifugation and filtration) give slightly different results: a set of measurements conducted by DSEA on 18 samples of programme no. 3 (30 November-1 December 1992) gives the following statistical values of the difference:

(SS by filtration − SS by centrifugation)

$$\text{mean} \quad = \quad +47 \text{ mg/l}$$
$$\text{standard deviation} \quad = \quad 52 \text{ mg/l}$$

Thus it appears that the results obtained by centrifgation are systematically less than those obtained by filtration (15 measurements by centrifugation show a deficit between 10 and 30 mg/l).

2.3.3 Analysis of data

For each of the following parameters—conductivity, total COD, particulate COD—two curves are shown (Figs. 14-21). The curves titled 'Intercalibration' reproduce the set of analytical data of ADES and laboratory data as a function of time. The experimental points in the four sets are joined by a solid line. The curves titled 'Correlation' enable visual assessment of the dispersion of analytical results.

Examination of the temporal variation of conductivity (Fig. 14) reveals that the concentration of dissolved mineral matter in waste water of work XIV is relatively constant in the dry season. On the other hand, in the drain at Brussels in which one can clearly perceive a daily periodic component related to the influence of a large flow rate of parasitic water (conductivity varying from 1 to 1.6 mS cm^{-1}), the variation of conductivity in the dry season on work XIV is only about 20% over a period of 24 hours. In the rainy season (see the results recorded in programme no. 9), the introduction of run-off entrains a significant reduction in conductivity.

For determination of the dissolved COD, the following hypotheses were used:

— The dissolved COD in the dry season is linearly related to conductivity, viz.

Dissolved COD (mg O$_2$/l) = K × conductivity (mS·cm^{-1})

The coefficient K is assumed to be equal to the ratio of the daily mean of the dissolved COD (mean of laboratory measurements: 132 mg O$_2$/l) and daily mean of conductivity (1.15 mS cm^{-1}); thus K = 114.

We wish to emphasise the fact that this simple formulation could not be used in the case of large contribution from a flow of parasitic water of non-null salinity. In this case, respective contributions of waste water and parasitic water to the overall salinity must be taken into account and this requires a knowledge of the flow rate of each component.

— In the rainy season, the complementary supply of dissolved mineral organic matter by the flow of rainstorm run-off is assumed to be negligible. It is thought that rain-water has zero salinity and that the pollutant added by overland flow is essentially of the particulate type. Such a hypothesis is not valid in certain circumstances, especially when salt is sprinkled on snow.

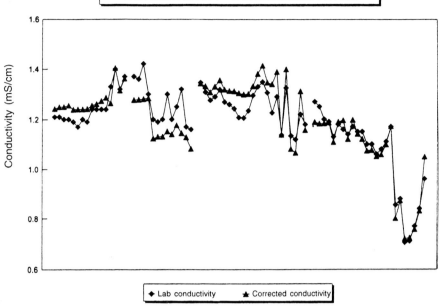

Fig. 14.

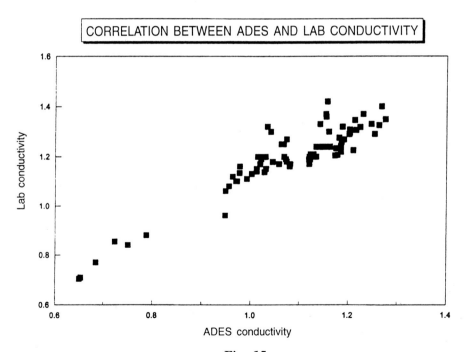

Fig. 15.

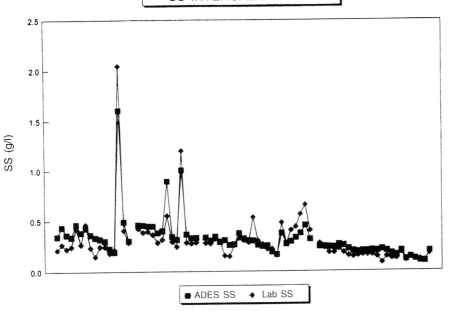

Fig. 16.

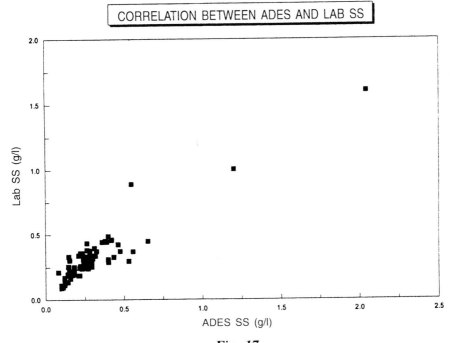

Fig. 17.

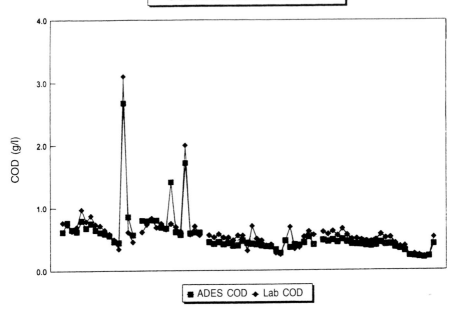

Fig. 18.

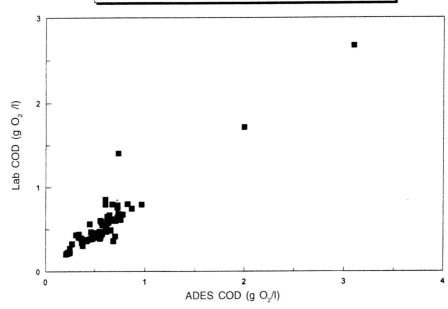

Fig. 19.

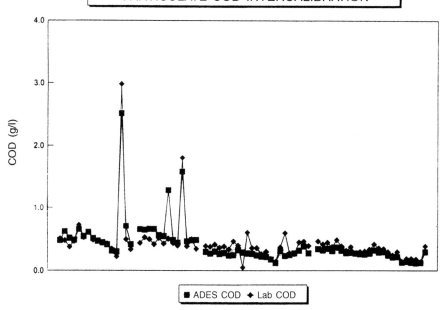

Fig. 20.

CORRELATION BETWEEN ADES AND LAB PARTICULATE COD

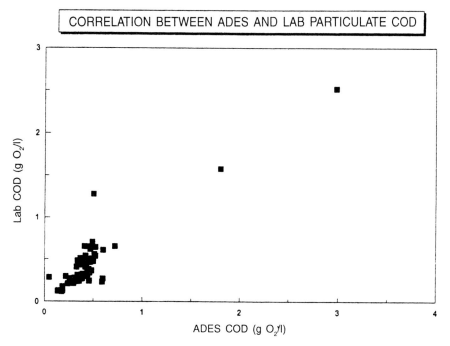

Fig. 21.

One may follow the above statistical approach to evaluate again the influence of the sampling procedure by comparing the value of the coefficients of correlation for conductivity and other quantities:

conductivity:	r = 0.90
SS:	r = 0.91
Particulate COD:	r = 0.90
Total COD:	r = 0.93

The values of the coefficient of correlation on the site of Creteil and the operative mode of DSEA, make it possible on the one hand not to reject the set of hypotheses of calculation and, on the other, to confirm that the dispersion of results is essentially due to the sampling procedure.

3. MANAGEMENT OF ADES STATION

3.1 STANDARD SUPERVISION AND MAINTENANCE

The various operations constituting standard maintenace of the ADES analyser are simple and require no special tools.

Frequency of maintenance is once a month.

Maintenance of the instrument falls into five major operations:
— visual inspection of the instrument,
— inspection and cleaning of gauges,
— recharging liquids (control and cleaning fluids),
— checking the operation of the compressor of the vacuum pump and pneumatic circuit,
— maintenance of a complete cycle of sampling and measurements.

Implementing the above takes about an hour.

The ADES analyser is provided with the material and software necessary to establish a two-way communication with a supervisory station, either local (with a portable terminal) or at a distance, by telephonic means (with a computer and a modem).

Communication with the ADES station located on work XIV can be established either through the Command Station VALERIE (Val-de-Marne informatic operation and regulation of effluents) or through the supervisory station of the service EA08.

This communication enables three operations:
— teletransmission of data stored in memory by ADES,
— telesurveillance of the analyser,
— telecommand of the analyser from the terminal.

Telesurveillance enables:

— surveillance in real time of the functioning of the instrument by means of a 'display board' capturing all necessary indications of actions in progress, signals dispatched by gauges, operational faults etc.,

— projection on the screen of data stored in memory by ADES,

— consolidation of various parameters of operation needed by the user.

The teletransmission includes transmission of measurements stored in memory by the ADES analyser as well as results of calibration and indication of operational faults.

The telecommand controls the dispatch of messages to the instrument for modifying the value(s) of one or several parameters (say, changing the interval between measurements).

3.2 PROBLEMS ENCOUNTERED

Regularity of communication with the instrument is vital for ensuring detection of all anomalies of operation and taking remedial action before the situation is aggravated.

Experience has shown that telesurveillance should be done on a daily basis; the manoeuvre takes just a few minutes.

Some faults related to blockage by a bend upstream of the inlet valve were remedied by modification of suction tubes. These obstructions were due to the particular functioning of work XIV which regularly receives 'foul' discharge from trucks of societies responsible for flushing (see Fig. 22, Sec. 3.3).

The most frequently encountered problem is telephonic communication. Actually, the instrument functions very well on the ground; it is during data transfer that interferences get introduced and sometimes result in a breakdown, necessitating reinitialisation. Ten such reinitialisations have had to be carried out since the instrument was installed.

In two years of operation, Aquarius has had to replace the compressor, valves of calibration and the cleaning fluid (once each) and the lower valve (twice). The electronic module had twice to be sent to the workshop (software problem, rewiring after two erratic false contacts).

The teletransmission of data is a simple operation but like all continuous measurements or continuous follow-up, ADES rapidly generates a mass of important data which has to be validated, interpreted and stored.

At present, data are analysed with the software 'EXCEL' which generates tables of values and the corresponding curves.

Recent linking of the ADES station of work XIV with the VALERIE system will enable visualisation and storage of this system. It will also enable sounding of alarms to signal operational faults at the measurement station.

3.3 TYPICAL DISPLAYS

The curves depicted in Fig. 22 are a typical display of the concentration of SS and COD for a period of one week. The particular functioning of work XIV makes it possible to distinguish the weekend, which is generally quieter than weekdays and relatively free from the activities of the cleaning team of the sewerage system, which regularly removes rubbish at work XIV and upstream of it.

Figure 23 describes the storm event of 21-22 January 1995 when the peak flow rate was around 1400 l/s. Reduction in the value of conductivity is very large: 1200 mS cm^{-1} before the storm, minimum around 0.1 mS cm^{-1}. Reduction in total COD is larger than that in SS and these concentrations show similar trends during the peak flow in the work.

3.4 INITIAL AND OPERATIONAL COSTS

The investment for the ADES station was nearly 250,000 F (1994 price) without value added tax (200,000 for the instrument and 46,000 for the software and teletransmission). To this must be added the cost of instrument installation (measurement chamber constructed as protection against urban vandalism), electricity and water connections.

The operational cost is mainly due to the maintenance contract (55,000 F/y). This amount may appear rather exorbitant but DSEA hopes that when more instruments are installed—the one acquired by DSEA was the first of its kind—this cost will reduce (another three instruments are to be purchased for the basin of Vitry-sur-Seine) and the firm Aquarius may be persuaded to open a maintenance service near the Parisian region.

At present, one needs to add the usual overhead (EDF, water, telephone) and personnel expenses for daily (a few minutes) and monthly (1 hour) inspection, validation and interpretation of measurements as per requirement. Just as for any other instrument, the expenses for initial intercalibration with laboratory measurements and renewal of operation from time to time (once every year) must also be considered.

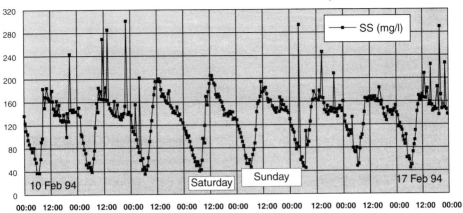

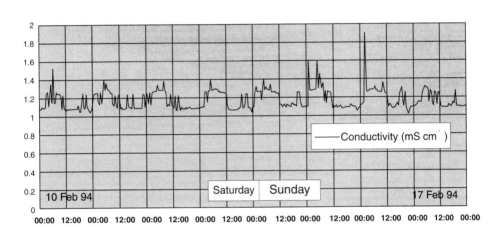

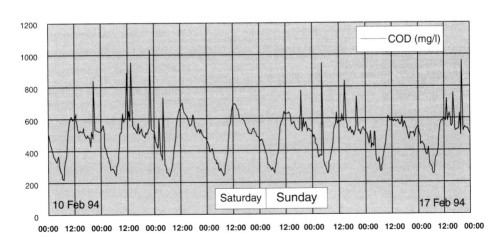

Fig. 22.

235

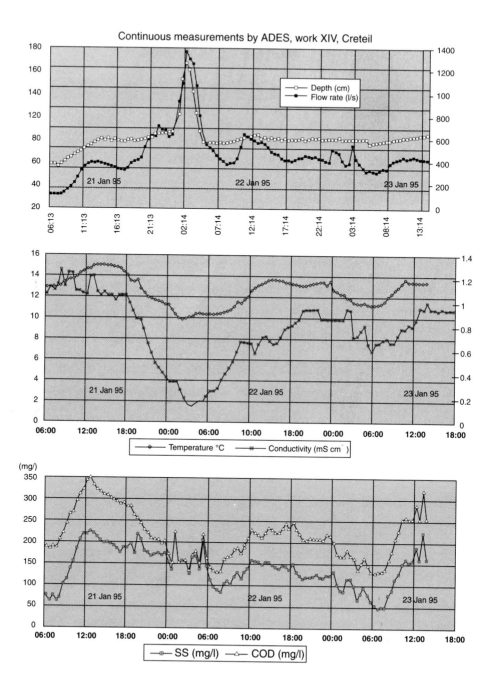

Fig. 23.

4. ADES AND MANAGEMENT OF THE BASIN OF VITRY-SUR-SEINE

4.1 BASIN DESCRIPTION

4.1.1 Introduction

The underground retention basin of rainstorm run-off, EV3, is situated under Jean Martin place at Vitry-sur-Seine. It receives the effluents of the Villejuif drain as well as the overflow of the combined sewer Paul Vaillant Couturier.

The Villejuif drain provides relief during the rainy season to the combined network as follows[1]:
— 4 chambers of diversion at the level of the works in Bièvre,
— interception of rainstorm run-off of the ZAC of Hautes Bruyères,
— 2 chambers of diversion of the combined works at the level RN7,
— 5 chambers of diversion of the combined works at the level RN305,
— 2 chambers of diversion of the combined works in the centre of Vitry-sur-Seine.

a) Goals
The downstream part of Villejuif drain, which paradoxically has the lowest hydraulic capacity, is rapidly saturated in the rainy season.

The basin EV3 was constructed to meet the following requirements:
1) offset inundations,
2) remediation of run-off from the Villejuif drain as well as the overflow from the Paul Vaillant Couturier sewer.

The volume of the effluent of the Villejuif drain is larger than that of the P.V. Couturier sewer.

EV3 ensures protection of the point of water intake from a location downstream (for the Ivry plant of potable water SAGEP) and the biological equilibrium of the Seine. The following operations are carried out in this basin:
— removal of flotsam,
— decantation of SS.

b) *Description*
The work is equipped with[2]:
— a pretreatment channel (screening, degritting and grease removal);

---

[1] See colour plate no. 1.
[2] See colour plate no. 2.

— 2 basins, B11 and B12 (18,500 m³) which absorb the first storm impacts and decant the SS;
— an upper basin, B2 (10,200 m³), which stores water under gravity;
— an EP tank for holding water by pumping.

In the dry season (Q < 0.5 m³/s) water flows only in the pretreatment channel. In the rainy season (Q > 0.5 m³) part of the effluents of Villejuif drain overflows in the basins.

4.1.2 Regulation

The duties of surveillance and management to be performed by personnel responsible for local supervision are detailed below.

a) *Filling (Impoundment)*
The various chambers of the EV3 are filled one after the other by successive overflow, depending on the volume of the inflow hydrograph, in the following order: B11, B12, BR, B2 and B3.

The automatic arrangement does not take into account the qualitative criterion for filling.

b) *Evacuation*
This is carried out in two phases:
— evacuation by gravity,
— evacuation by pumping.
The procedure is as follows:
— sludge pumps are operated during every evacuation of basins and cleaning of B11;
— higher parts of B11 and B12 are evacuated by gravity and lower parts by pumping;
— B2 is evacuated by gravity;
— B3 is evacuated by pumping.

c) *Cleaning*
Basin 11 is cleaned by a set of three buckets, basin 12 by scrapers and basins B2 and B3 manually.

d) *Treatment of sludge and wastes*
This operation is carried out using the following equipment placed above the pretreatment channel: screens, compacter and hydrocyclone. The dirty water from the hydrocyclone is discharged towards the combined network and diverted to the treatment works. The water containing sand is directed towards B11.

4.2 UTILISATION OF ADES MEASUREMENT STATION

4.2.1 Location

Measurement of the quality of effluents is conducted at three points:
— inlet to the basin in the Villejuif drain,
— in basin B11,
— in the holding channel.

The height of suction varies from one measurement site to another.

The suction tube is placed on the floor of the inlet to the basin and the holding canal.

The suction point level (9 m) in the basin necessitates installation of a high-powered pump.

4.2.2 Commencement of evacuation procedure

In the automatic system currently in use, commencement of the evacuation procedure is governed by three criteria:
1) The water level downstream of the drain should be lower than the prescribed value.
2) The concentration of SS should be lower than the prescribed value or the rate of reduction of SS should be faster than the prescribed value; these values are prescribed by the ADES station in basin 11.
3) A temporisation t corresponding to a time of decantation must have elapsed.

As there are several pumps in the EP tank it is possible to modulate and accelerate the speed of pumping if the automatic system detects that the level in the drain upstream of EV3 has risen again.

The basin will be handed over to the Department of Val-de-Marne by the end of July 1995.

Management of the basin involves the following considerations:
— regulation of fixed thresholds for filling various compartments;
— value prescribed in connection with commencement of evacuation;
— putting the sludge pumps into operation right at the start of evacuation.

It may be noted that at present only the ADES station at B11 is used effectively for qualitative management of the Vitry-sur-Seine basin; the other ADES station merely maintains an inflow-outflow balance.

The DSEA has prepared a programme for further study of rainstorm run-off in the catchment area and its treatment in the Vitry basin. These studies ought also to find a solution to the problems mentioned above and eventually lead to improvement of the automatic system currently in use.

239

CONCLUSION

Since the end of 1992 DSEA has used the ADES measurement station as a 'watchdog' on the quality of effluents of waste water transiting through work XIV at Creteil.

Various programmes of intercalibration of measurements conducted with ADES and laboratory measurements have yielded a coefficient of correlation value of about 0.9 for the parameters of interest (conductivity, SS, particulates and total COD).

These results were obtained through the good facilities of management and the special features of the instrument. This has encouraged DSEA to install three more ADES in the basin for purification and storage of rainstorm run-off at Vitry-sur-Seine. These instruments will be very helpful in acquiring better knowledge of the Villejuif drain and improving treatment in the various compartments of the basin. They will also prove helpful in planning evacuation of held water and the overall efficiency of the system.

Based on the values obtained from ADES it is felt that the VALERIE system of DSEA is now capable of managing the flux not only in terms of quantity, but also qualitatively.

Lastly, this type of instrument together with a device for measurement of flow rate may provide a solution that meets the exigencies of recent regulations requiring determination of the volume of overflow to the natural environment in the rainy season from the largest storm weirs.

DISCUSSION

Mr. Valiron: *Is the **calibration** of ADES carried out at Creteil always valid at Vitry, bearing in mind that the effluents are different?*

Can one be sure of obtaining correct responses when one compares the three instruments in the Vitry cascade? Is not a brief complementary calibration necessary?

Mr. Rabier: *Calibration of the three instruments at Vitry is in progress.*

Each instrument installed is presently being subjected to specific calibration.

Mr. Van der Boght: *Every time an instrument is put into operation, the initial calibration is used and on-site intercalibration carried out. To date, verification of calibration of these new sites has necessitated no modification of the parameters of the initial calibration.*

Mr. Gousailles: *SIAAP is currently negotiating with Aquarius the purchase of ADES. The instrument appears to be very reliable and enables simultaneous determination of the total, dissolved and decanted COD. It gives a rapid, continuous and pertinent assessment of the manner in which the components of polluted water will react to the systems of decantation.*

SIAPP is planning to install this equipment at Clichy.

CASE NO. 5

Summary of Surveys of Weirs

F. Valiron

This summary reports the main results of three visits to the four districts 75, 92, 93, 94 and SIAAP. It also highlights the steps needed for improving the management and efficiency of the 'complex network' comprising the installations of these five administrative units.

1. WEIRS AND BYPASSES

1.1 BRIEF INVENTORY

Slightly more than 400 works of discharge to the natural environment exist in the four districts of Paris and the Premiere Couronne; they are listed in Table 1.

About 60 of these works discharge into the drains of SIAAP; the water discharged ultimately reaches the Seine or Marne directly in the case of 22 works and via overflow channels, often long, for the remaining 38.

Most of these works are weirs performing various functions. Some are bypasses enabling passage of the flux in one sewer to another. There are also main sewers carrying run-off but most of them are not included in the list of Table 1. The functions performed by these works are described in Table 2.

Only the works of cases 1, 2 and 3 corresponding to 'weirs in the strict sense' and of a direct discharge type are covered by legislation.

In addition to these weirs there is a large number of bypasses and other similar works interlinking two main sewers (cases 4 and 5; more than 400 in number).

Thus there are 800 works, weirs, sewers of run-off and bypasses playing a role in discharge to the natural environment during storm episodes (and sometimes during the dry season). The number for the ensemble of the Parisian agglomeration would be of the order of 1000 if one includes the discharge from Val d'Oise, Yvelines, Essonne and marginally from the Seine and Marne.

	Number of works	Points of discharge
Seine St Denis	85	69 to Seine-Vieille Mère-Morée 16 to Marne
Hauts de Seine	91 progressively reduced to 36 2 Ru de Marivel Ru de Vaucresson discharges of 11 communes and 25 weirs carried to SIAAP to Seine	to Seine RG 63 left bank 28 right bank to sewer SIAAP doubly South to Bièvre and Seine via SIAAP
Paris	43	to Seine at Paris or in districts 92 and 94
Val de Marne	30 40 } principal 20 and 80 others, of which 40 are departmental	to Seine to Marne to Yerres
Total	416	

Table 1

At present there is no proper register of all these works. This incomplete record of weirs in the strict sense will be rectified within the year. In the case of bypasses, however, data are lacking, at least with respect to their dimensions and consequently their impact. Part of a study in progress is devoted to acquiring better knowledge of the networks but will take a long time and involve large costs.

Lastly, about 70 pumping stations located near some weirs lift water to the Seine or Marne in case of flood. Their capacity has to be enhanced at several points to prevent overflow of sewers. A sufficiently large number of weirs, especially some in Paris, require installation of mobile pumps when heavy rainfall occurs.

The network of sewers of four control districts of the Ile of France is spread over 760 km², of which 510 sewers are of the combined type. The network serves a population of six million inhabitants. It also receives waste water and run-off from exterior zones spread over more than 1000 km² and supporting about 2 million inhabitants. At present, the collected water flows to three treatment

1	Discharge of a combined sewer during rainy season	to the River Seine
		            Marne
		            Yerres
		            Morbas-Morée
		            Vielle Mère
		
		Junction with SIAAP drain and maybe/maybe not discharge to Seine or Marne

2	Discharge of a combined sewer to storm sewer	Ends in Seine-Marne or often in a tributary via retention basin
3	Discharge of storm sewer to the natural environment	to Seine, Marne etc.
4	Discharge of a combined sewer to another combined sewer	within network
5	Combined sewer directing water of dry season to a waste-water sewer	Bypass (examples, flow of Valenton)
6	Rain-water baffle directing water of dry season to waste-water sewer (or retention basin)	Bypass

Table 2

works—Achères, Valenton and Noisy, while two are under construction—Colombes and Bonneuil (Fig. 1).

The map in Fig. 2 shows the effects of discharge from weirs during a heavy storm on the water quality of the Seine or Marne.

Dégardin [44] used the FLUPOL model to calculate the quantity and quality of overflow during a decennial rainfall assumed uniform and determined the zones of discharge with the heaviest concentration of SS. He also calculated the volume of inflow which helps in carrying out satisfactory calibration of the model and in checking measurements of pollution brought in at some points.

1.2 Types of weirs and their equipment

The number of weirs is large and they are of many varieties; some are old and others more or less recent. Differences also arise from site constraints.

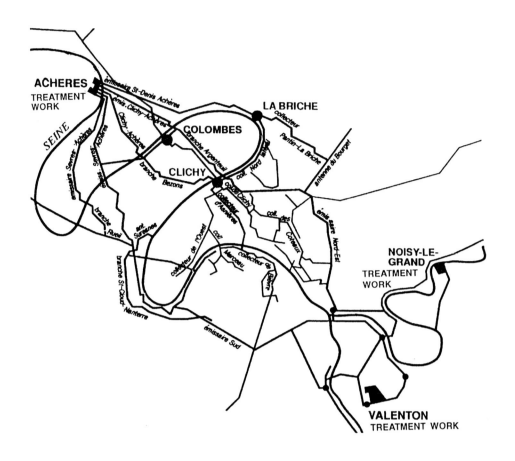

Fig. 1: *Treatment works and the zones served*

Up to now, their only function has been the hydraulic diversion of flow exceeding the downstream capacity, in order to preclude overflow. Barring some weirs constructed very recently, no weir was designed to direct maximum pollution to the treatment works. Therefore, no weir has an attached settling chamber and very few are equipped with screens to prevent the flow of flotsam to the river. Most have adjustable shutters and are protected against floods by movable dams manually installed. Today, fewer than 60, say about 15%, are equipped with adjustable valves that can be manipulated by remote control and protected against floods by automatic or remotely controlled flaps. This situation will change very soon as most of the large weirs have been or are being modernised.

To reduce the unaesthetic aspect of flotsam in the river, SIAAP constructed eight barrages in the zones of accumulation of wastes on the Seine and Marne. The breadth of the barrages was kept small so that there is no interference with navigation. A special boat collects 750 tons of such matter and

Fig. 2: *Effects on natural environment of discharge from weirs during a heavy storm.*

transports it to a site where it is destroyed. When 12 works have been constructed—in 1995—1500 tons of flotsam will be removed, of which 30% is brought by the sewer[1].

Lastly, the present means of continuous measurement of flow rate are insufficient; the data collected are transmitted to a central monitoring station where the decree of March 1994 makes immediate analysis of it obligatory.

1.3 BYPASSES

These, too, are of many varieties—from the type providing a simple connection between two sewers with fixed thresholds, with/without the possibility of a

---

[1] See colour plate no. 14.

regulating mobile dam, to that resembling a weir with adjustable valve, or siphons with automatic priming mechanism, or even pumping stations. Most of these works are simple connections, often unnoticed even by the users; it is only during an inspection in response to unusual malfunctioning of the network that one becomes aware of their existence. Nevertheless, they play an important role because they strongly modify the sharing of feeding basins on a permanent basis or when the flow rate exceeds a certain prescribed value.

The skimming done by these works through removal of water of the dry season flow to combined sewers can change the classification of a main sewer from a combined type to a quasi-storm run-off type. Knowledge of these works, their effects, their equipment (such as valves or devices for remote regulation) is thus essential for improving management, all the more so if changeover from a static to a dynamic management is desired.

1.4 MEASUREMENTS ON WEIRS

Until ten years ago measurements were rarely conducted. Today, however, they are regularly done, primarily to determine the amount of water discharge, but also to determine the pollutant load.

Some measurements are conducted simultaneously on several weirs in a series and take into account exterior inflow, though not in a detailed manner. Surveys have to be conducted to accurately determine the topography of works which are old or whose description does not exist. At present, no use is made of the records to establish the rate of impermeability and the inflow during rainfall corresponding to the flood under study.

Measurement of pollution is conducted only on the overflow and none on the load withheld by the work. The quantities measured are SS, BOD, COD, sometimes nitrogen but almost never the non-decantable dissolved organic pollution.

Lastly, to evaluate the impact of discharge, measurements should be simultaneously conducted on the river also, something not done to date. A plan is under consideration with respect to the circuit of Boulogne in liaison with a measurement station on the Seine, of the network 'APES' of Hauts de Seine (see case study no. 6).

Without doubt, these lacunae should be removed, at least at certain selected points, to obtain a better overall picture. And such is feasible because there are about twenty measurement stations on the rivers in the Parisian region, continually analysing most of the pollutants (Fig. 3).

Some measurement stations are organised as networks and are managed by a software (CALYSTO). All the data are regrouped in the 'Regional Laboratory of the Quality of Rivers of Ile de France'. Utilisation of the data of the rainy season together with that of the storm weirs could prove fructuous: values of

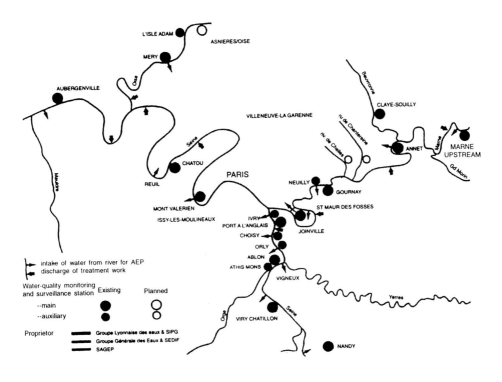

Fig. 3: *Water-quality measurement stations*
on the rivers of Ile de France.

the flow rate of rivers during storm episodes, at least for some sites, could definitely be derived from this information.

Measurements of pollution conducted at Val de Marne on water of the dry season from drains directed to Valenton, are useful. Undoubtedly, a detailed analysis of the inflow would be profitable, if only to ascertain the presence of certain anomalies in the collection of these works.

It may be noted that these measurements have a double objective:

— Obtaining 'depth-flow rate' curves of fixed or adjustable weirs and curves of correlation between flux discharged and measurement by a turbidimeter or even calibration of a model.

— Providing better knowledge of the amount of discharge or polluting flux for an annual assessment or assessment during storm events, or even modifying the position of one or several valves on the basis of results of continuous measurements.

Measurements of pollution at the stations located in Seine St Denis and Val de Marne give rise to the hope that in the near future management will include consideration of this aspect.

Automation of more and more weirs and pumping stations and linking them to a control system opens the way to assessments via stored data. But due to lack of personnel, very little progress has been made in this direction.

2. TANKS ASSOCIATED WITH WEIRS

Although a number of 'storage-treatment' tanks for reducing discharged pollution have been planned, to date not one has been constructed. Taking advantage of the works of the 'Grand Stadium', the first such large tank (160,000 m³) is to be constructed for treatment of part of the discharge of the Briche weir. In total, more than 20 large tanks with a cumulative capacity of more than 1 million m³ need to be constructed to obtain a reduction of 70% in the pollution discharged during rainfall of the annual return period. The sites earmarked for these largest tanks are plotted in Fig. 4. On those which pose problems of terrain the tanks could be replaced by galleries, also serving as interconnections between drains. SIAAP has already undertaken a study of the situation.

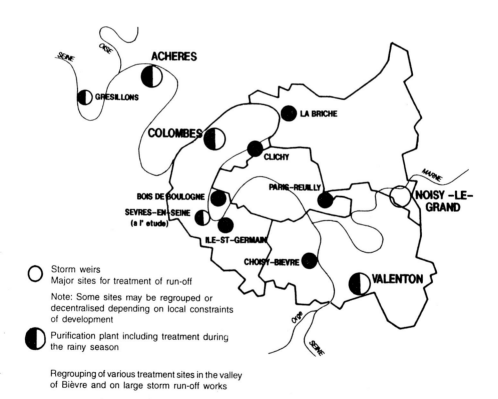

Fig. 4: *Major basins of 'storage-treatment' associated with weirs.*

It has been shown that while calculating the capacity of a tank for a weir, due attention should be paid to measurement of flow rate and flux; for large catchment areas, use of typical rainfall leads to much lower values.

Recent studies have demonstrated that for some tanks it is advantageous to combine natural decantation with a lamellar decantation with flocculation for part of the flow rate.

On the other hand, a large number of 'laminated storage' tanks were constructed; their capacity was about 1,500,000 m³. They have some effect on reducing pollution but that depends on managerial procedures. Construction of several such tanks is planned in the years to come to fulfil the objective of remediation and maximum detention time for stored water.

3. ZONES OF COLLECTION OF MAJOR WEIRS

The map in Fig. 5 shows that a number of collection zones of the same work are located in more than one district and/or belong to several proprietors.

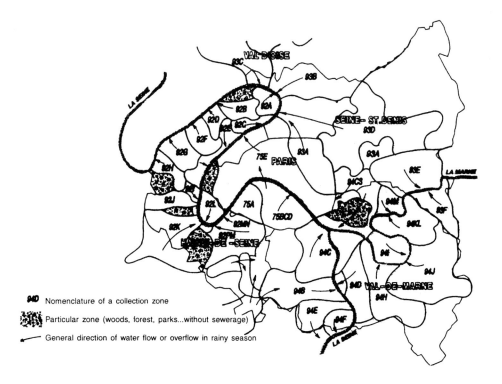

Fig. 5: *Service basins of about forty points of discharge in the Seine and Marne rivers (Dégardin [44]).*

For example, analysis of discharge of sector 75D (Fig. 6) and sector 75C (Fig. 7) demonstrates the imbricating effects of various weirs on the inlet to the South drain of SIAAP and the inflow from two districts (75 and 94) and SIAAP.

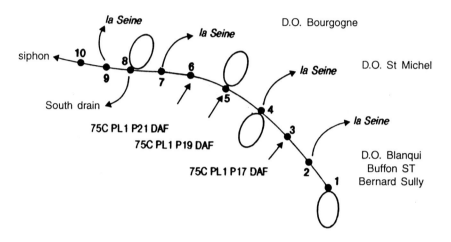

Fig. 6: *Bas sewer of Paris (75D).*

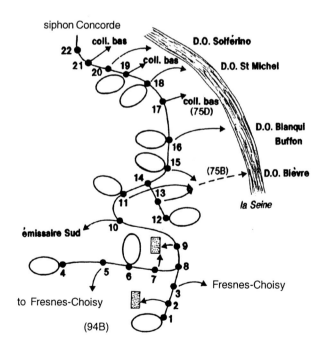

Fig. 7: *Bièvre and Paris right bank (75C).*

This overlapping, at least between SIAAP and a district and often between two or among three districts, requires closely co-ordinated management to take full advantage of storage in the upstream network. But the present situation is very different: a connection between informatic systems exists only between SIAAP and Seine St Denis and even this is limited to exchange of information (via Partenaire) though fortunately in real time (see case study no. 7, Sec. 4).

Lastly, it may be noted that studies for management of weirs were conducted (and some are currently in progress) in each district. Their utility is indeed uncontestable but would it not be enhanced were they conducted on an overall basis or more simply jointly by the feeding basins?

Results of **elaboration of the directive plan of Bièvre** which associated districts 92 and 94 with SIAAP, the first study of a complex basin to be initiated (described in Figs. 6 and 7), show that this is indeed a **useful direction to pursue.**

This plan is part of an analysis of a number of studies and measurements conducted by three proprietors and models of the operation of this complex network and its three main outlets: Fresnes-Choisy and Villejuif sewer in the east, Bièvre-Bas and its weirs in the north, and the two southern drains in the west (Fig. 8).

The various scenarios examined show the **preponderant influence** on the flux in terms of quantity, quality and distribution, for the treatment at Achères and discharge to the Seine of **some 70 works or bypasses.** They also indicate that it is necessary to equip about ten of them with control devices. It is further necessary to clearly define the objectives suggested by the three proprietors:

— priority of directing to Achères the maximum flow containing as much waste water as possible and of getting rid of the maximum quantity of parasitic water;

— reduction of inundation due to local insufficiency of means of transport;

— reduction of storm run-off pollution discharged in the Seine.

Figure 8 shows the distribution of the total quantum of SS currently discharged in the Seine; results are summarised in Table 3.

The hourly flux is close to the mean hourly flux of the dry season which reaches the Seine without treatment and is of the order of 4 tons/h. This demonstrates the large drainage caused by storm events while the annual pollution attributable to rain represents only about 3% of the annual contamination of the dry season.

For the annual rainfall, 40% of this discharge comes from Fresnes-Choisy, 28% from the Parisian storm weirs including Watt, 20% from the South drain and its storm weirs, and 12% from Issy les Moulineaux.

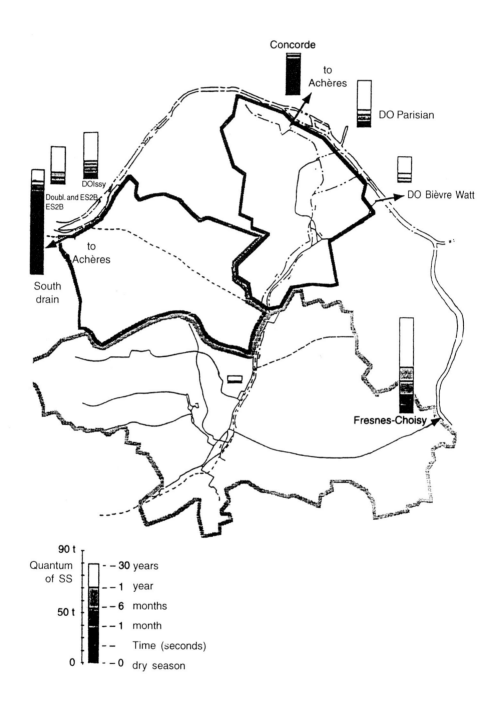

Fig. 8: *Distribution of the quantum of SS in different outlets of Bièvre basin during the storm events examined.*

	Monthly rainfall	Six-monthly rainfall	Annual rainfall	Three-year rainfall
Total	25 tons	50 tons	80 tons	200 tons
Hourly flux	5.5 tons	12.5 tons	19 tons	50 tons

Table 3: *Total flux discharged (in tons)*

Reduction of discharge of pollutants will require construction of 180,000 m³ of basins for 'storage and decantation' situated at five sites and overflow of about 350,000 m for 'laminated storage'. The effect of these works on an annual rainfall would be a reduction of 90% in the pollution discharged (according to the plan) but this reduction may be based on a gross overestimation of the actual efficiency of works, which at present is not well known due to lack of experience.

4. MANAGEMENT OF WORKS

Three types of management can be distinguished.

4.1 STATIC FROZEN MANAGEMENT

That in which monitoring devices are set once and for all and due to lack of motivation and repair shops, the hydraulic flow and contaminating flux are not modified over time; discharge of the rainy season may increase due to lack of maintenance causing overload in the water channels. This is yesterday's management!

4.2 PROGRESSIVE STATIC MANAGEMENT

Making use of better knowledge of the network and its reactions as well as of the indications of flow models and study of scenarios, one can, in this type of management, modify the calibration of works in view of priority objectives or even remedy malfunctions by constructing new works (sewers, basins, screens etc.). Such modifications or adjustments are not carried out during storm episodes and thus do not take into account variations of flow rate or water depth occurring during such episodes. Improvements necessitate:
— Knowledge of the geometry of the network and sufficiently accurate measurements for calibrating the model. Measurements should be conducted on the basin to be improved and, if necessary, extended beyond the administrative limits.

— Dialogue among the participants to arrive at an understanding regarding the objectives and priorities.

It may be noted that knowledge of the effects of rainfall on the network and analysis of malfunctions are easier for local overflows than for disorders caused by the river. This explains why finer adjustments are done first to remove local faults.

4.3 DYNAMIC MANAGEMENT

This differs from the preceding type in that adjustments of works are carried out during the events to be controlled. In addition to the conditions listed in Sec. 4.2 above, management requires:

— modifications of works by remote control, either directly or by adjusting the command device even during the crisis;

— continuous collection of management data with essentials transmitted to the central control station.

A study conducted at Bièvre (see Sec. 3) revealed that there are marked advantages of this type of management over the one described in Sec. 4.2, even when applied to the same network and the same complementary works: 10 to 20% less volume of discharge, at least for a decennial rainfall, significant reduction in pollution and the residual stored volume at the end of a storm, though slightly more, can be entirely treated at Achères.

This justifies the efforts made for changeover from progressive static management to a dynamic management. The changeover is effected by better management in the districts, improvement of equipment used for measurements and introduction of remote control. Dynamic management has already been introduced in some parts of the network, for example in drains through the use of SCORE and in some storage basin outlets in Seine St Denis.

Equipping weirs with adjustable valves is also part of dynamic management; the discharge is reduced by means of storage in the upstream section of the sewer and remediation is ultimately augmented by installation of devices of continuous measurement of pollution and use of software for modification.

Extension to sectors which are more widely spread out would be highly beneficial but would take a long time. It requires:

— a more serious dialogue among participants to reach an agreement regarding the exact goals of management;

— assessment of discharge, both its volume and quantum of pollutants, for the main zones, annual as well as during various rainfalls, in order to make proper arrangements for evaluating and measuring the progress;

— closer co-ordination among the various informatic systems of management.

Every manager should feel that it is worth spending time and effort to enhance knowledge of his network, analysing digital data stored in the memory of his system of management and exchanging information with other managers, if the goal of more effective management is to be achieved. Some steps have already been taken in this direction, for example the study of Bièvre's contribution.

It may also be noted that one of the fallouts of modelling the network is knowledge of its storage capacity during the rainy season (volume stored or volume that can be stored in excess of the dry season limit) and the means of increasing it to take full advantage of the potential capacity (bypass, valves in network etc.). Considering the competitive nature of costs, this measure would help to reduce the number of new works of laminated storage or storage-decantation to be constructed.

The attention of users of the networks should be drawn to the fact that **'reactions' of an automated network,** just as for every work concerned, differ from those of a classical network. Special provision should therefore be made for the **safety of personnel performing special duties** and the possibility of slowdown of activity in the works during maintenance. Such provisions have already been made at Seine St Denis and are in the course of installation at Val de Marne.

CASE NO. 6

Management of Networks and Storm Weirs of the Parisian Agglomeration

M. Soulier, DEA Hauts-de-Seine

1. INTRODUCTION TO ZONE STUDIED

The catchment area of the Boulogne loop comprises the city of Boulogne, area 550 ha, on the right bank and parts of the communes of Issy-les-Moulineaux, Meudon, Sèvres, Saint-Cloud and Suresnes with a total area of about 950 ha on the left bank (Fig. 1).

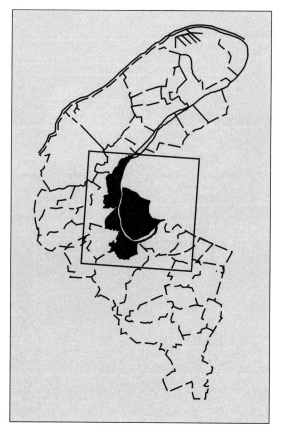

Fig. 1.

The sewerage network is of the combined type and the effluents converge towards two banked main sewers paralleling the Seine at the exit from Paris up to Suresnes bridge. The effluents of the dry season and light rainfall rejoin, at Suresnes bridge, the outpost Suresnes of SIAAP.

1.1 NETWORK OPERATION UP TO 1960s

1.1.1 Right bank of Seine (Fig. 2)

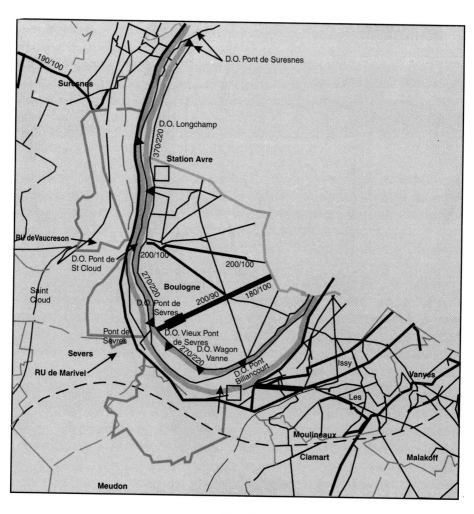

Fig. 2.

— During the dry season and light rainfall, water is evacuated up to Suresnes bridge, crosses the Seine through a siphon and rejoins the outpost Suresnes of SIAAP.

— During heavy rainfall the excess water is discharged to the Seine by four storm weirs: 'Wagon-vanne', 'Vieux pont de Sèvres', 'ACBB' and 'Pont de Suresnes'. The weir at Pont de Suresnes is of the frontal type while the other three are lateral and of small dimensions (1 m long sill) and adjustable by means of shutters.

— During Seine flooding the weirs are closed and the rainstorm run-off is diverted back towards the Seine by the station 'Avre' located at Boulogne.

1.1.2 Left Bank of Seine

During the dry season and light rainfall, water is evacuated up to the Suresnes bridge where it rejoins the outpost Suresnes of SIAAP. The sewer along the left bank of the Seine is also fed by the group of rivulets of Marivel and Vaucresson, respectively at Sèvres bridge and Saint-Cloud bridge.

During heavy rains the excess water is discharged to the Seine by three storm weirs of the same type as on the right bank: 'Vaugirard', 'Longchamp' and 'Pont de Suresnes. Two other storm weirs evacuate the excess water of the rivulets of Marivel and Vaucresson.

During Seine flooding the three departmental storm weirs are closed and the rainstorm run-off is diverted back to the Seine by the station 'Vaugirard' located at Issy-les-Moulineaux. The topography of the site enables rainstorm run-off of the rivulets of Marivel and Vaucresson to be evacuated to the Seine by gravity.

1.2 PRESENT OPERATION OF THE NETWORK

The major modifications carried out are listed below.
• Linking the sewers along the right and left banks by a siphon under the Seine at Sèvres bridge.
• Hydrological studies suggest that the location of some storm weirs on the right bank of the Seine should be reviewed and their size increased. The state of weirs of the left left bank, upstream and downstream, is as follows:
— Billancourt bridge: weir under construction; centralised control.
— Wagon-vanne: weir unaltered; ultimately to be transferred to central control.
— Old bridge at Sèvres: to be abandoned soon.
— Sèvres bridge: at present under local control but soon to be transferred to central control.

— ACBB: to be abandoned soon.

— Suresnes bridge: weir modernised; at present under local control but soon to be transferred to central control.

The storm weirs of the left bank have not been altered thus far.

1.3 HYDROLOGICAL INSTRUMENTATION

The scope of the study includes installation of devices for measurement of rainfall and flow in the networks. Some instrumentation was carried out in the 1980s during installation of the system known today as 'GAIA' (Management supported by informatics of sewerage).

At present there are 3 rain gauges and 9 measurement stations (7 on the right bank and 2 on the left) equipped with 37 devices for measuring depth and 40 for measuring flow velocity.

2. INTRODUCTION TO HYDRAULIC SIMULATIONS

The Boulogne zone has the proper characteristics vis-à-vis the problems of reduction of discharge of contaminated water during the rainy season. In fact, most of the sewers can be inspected and the slopes are small (less than 1 mm/m); the right bank sector has a storage capacity of about 30,000 m³.

For better appreciation of the possibilities of reduction of discharge of residual water of the rainy season, the network was modelled by means of the software MOUSE (Modelling of Urban Sewers) and the model was calibrated through utilisation of the information collected by the system GIAA for hydraulics and measurement of rainfall.

Calibration of the model was carried out over three rainfalls of different characteristics: storm, long-duration rainfall of 'constant' intensity, and long-duration rainfall of 'constant' intensity followed by a brief rainfall of peak intensity. Simulations consisted of calculating the overflow from storm weirs with a fixed threshold followed by simulation of network operation by replacing the fixed thresholds with adjustable ones.

Thus for the rainfall of 4 December 1992 (protracted, followed by a peak), similar to a rainfall of the period of return close to 2 years, the overflow decreased from 41,500 to 21,800 m³, say a reduction of 50%.

3. CONSIDERATION OF QUALITATIVE PARAMETERS

To improve the management of works of flow regulation (storm weirs, valves, pumping stations) and envisioning a system for real time, the Department has undertaken studies in the qualitative domain for integration of the concept of impact of discharge of residual water on the natural environment.

The steps undertaken are described below.

3.1 STUDY OF POLLUTION FLUX OF THE RAINY SEASON FROM AN URBAN CATCHMENT AREA AT BOULOGNE (FIG. 3)

The objectives of this study were to ascertain the contribution of water and pollution by overland flow from an urban catchment area and to understand the change in quality of discharge in the network. The study was carried out in two phases:

— **Measurement phase:** Three points of measurement of flow rate located on the network and 3 points of measurement of quality on the gullets. Analysis covered physicochemical parameters and metals (turbidity, conductivity, pH, COD, NH_4^+, chloride, SS, POM and for some samples BOD_5, TOC, TKN, P, hydrocarbons, Pb, Cd, Zn).

— **Modelling phase:** Modelling of the quality of the sewerage network was accomplished using the software MOUSETRAP to simulate the accumulation and washing of pollutants at the surface of the catchment area, transport of these pollutants in the network, and chemical and biological changes.

The objectives of modelling were: (i) understanding the phenomena, (ii) prediction of impact and (iii) optimisation of investment for the construction of works of remediation.

The catchment area of Boulogne was selected as a test case in consultation with the Agence de l'Eau, which participated in the study. This catchment area is spread over 48 ha, of which 80% is impermeable and the length of the network is 4600 m.

This study enabled determination of pollutograms at different points of the network, in particular on storm weirs, and thus improvement of the simulations of water quality in the river; it was completed in early 1996.

3.2 EQUIPMENT AT WATER-QUALITY MEASUREMENT STATION AT BOULOGNE

The equipment should be capable of continuous measurement of the quality of effluents of the dry and wet seasons downstream of the commune of Boulogne and include a device for gathering the information needed to integrate qualitative aspects in the management of storm weirs. The equipment of measurement of the quality of effluents is installed at the flood station 'Avre' at Boulogne located at the fringe of the woods.[1]

---

[1] See colour plate no. 7.

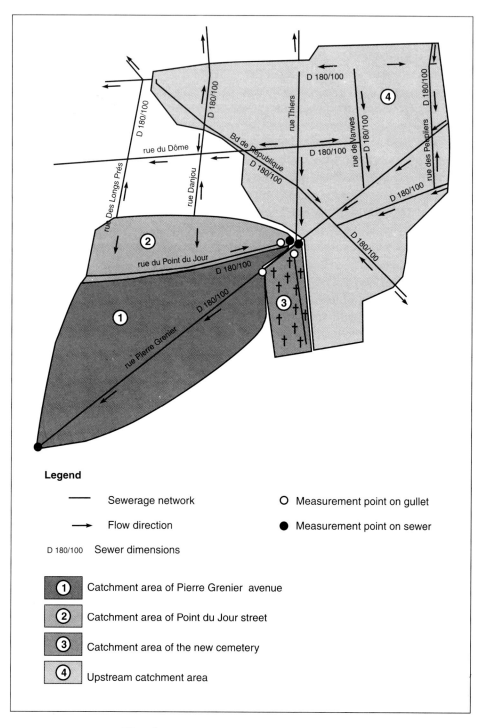

Fig. 3: *Catchment area of Boulogne.*

At the time of its erection the only function of the Avre station was to divert rainy season water of the right bank sewer during flooding of the Seine; now two modifications have been incorporated:

— Screens have been installed to strain water of the dry and wet seasons to improve, on the one hand, operation of the Suresnes siphon situated downstream and, on the other, reduce the flotsam discharged in the Seine during the rainy season at the Suresnes bridge weir downstream of the right bank sector.

— A water-quality measurement station has been installed in the pumping station downstream of the screens.

Technical problems arise from the fact that the crucial parameters of this impact (organic matter: BOD, suspended matter) are not readily measurable directly on the site (complex and expensive instruments are required) and there are no simple gauges that can provide direct continuous and reliable values. Given the present state of technology, one has to resort to indirect measurements based mostly on optical procedures; unfortunately these measurements do not always correlate well with the values required for crucial parameters.

To overcome these difficulties, it was decided that measurements should be taken on raw water, homogenised water (50 microns) and filtered water (0.5 microns) which contain only dissolved pollution.

The following parameters were measured:

— **raw water:** turbidity, temperature, dissolved oxygen, conductivity, pH;
— **homogenised water:** turbidity
— **filtered water:** conductivity, NH_4^+.

This group of measurements constitutes the 'primary' cycle. A 'secondary' cycle is linked with it which enables sequential measurement of BOD for each type of effluent and on samples in which time is often advanced by an adjustable step (markedly dependent on the season or type of storm event).

The BOD is measured in accordance with the norm AFNOR T90103, i.e., according to the method of dilution (dechlorinated water, saturated in O_2 and incubated at $20 \pm 0.5°C$). The consumption of oxygen is continuously recorded by gauges. The measurement may last 5 days (depending on the norm) or a shorter adjustable period. The BOD is computed by extrapolation using a correlation formula. The equipment comprises 12 measurement stations which permit a large number of combinations.

The system includes a data acquisition station connected by a modem with a microcomputer for processing information. Installation was completed in the summer of 1995.

3.3 MEASUREMENT OF QUALITY OF THE SEINE.
SYSTEM OF ALERT FOR POLLUTION OF SURFACE WATER (APES)

The objective of the system is continuous measurement of various parameters characterising the water quality of the Seine at different sites in the District. It provides information regarding the nature and location of accidental pollution and the global impact of activities carried out in the District.

The operation APES presently in progress (June 1995) was launched in collaboration with the Syndicat des Eaux of Presqu'Ile de Genevilliers, the Lyonnaise des Eaux, the firm Eau et Force and the Compagnie Générale des Eaux. From upstream to downstream, five measurement stations have been installed[1]:

— station of the Ile Saint Germain, large work, at Issy-les-Moulineaux, controlled by the district;
— station of Suresnes, right bank, controlled by the district;
— station of Villeneuve-la-Garenne, controlled by the district;
— station of Rueil-Malmaison, Marly chord, controlled by the district;
— station of Chatou (Yvelines), controlled by the Lyonnaise des Eaux (old AESN station).

As part of the study of the Boulogne loop, the first two stations delimit the part of the Seine on which residual water is discharged.

Measurements made at the various stations are given in the Table below[2].

	Rueil	Chatou	Villeneuve	Suresnes	Issy
pH	X	X	X	X	X
Oxygen	X	X	X	X	X
Temperature	X	X	X	X	X
Redox	X	X	X	X	X
Conductivity	X	X	X	X	X
Turbidity	X	X	X	X	X
Ammonium	X	X	X	X	X
Phosphates	X	X	X	X	X
Phenols			X		
DHIR	X		X		
UV			X		
Visible			X		
Cadmium	X		X		
Lead	X		X		
Copper	X		X		
Chromium	X		X		
Nickel	X		X		
Zinc	X		X		

[1.] See colour plate no. 4.
[2.] See colour plate no. 5.

In addition, the Suresnes station provides information on test 'fish' and once a day, the measured value of nitrates. Other quantities are measured every 20 minutes at every station. The information is collated in the office of Water at Mont Valérien, seat of the system[1], and transmitted to various real time supervisors, especially the District, through specialised France Télécom lines. The values are stored in a data bank located at Mont Valérien, also accessible through special lines.

The current investigations concern the multiparameter validation of information, the categories of quality of the Seine and transmission of mean values to the Regional Laboratory of Ile de France, in the format SANDRE.

3.4 COMBATING VISIBLE POLLUTION OF THE SEINE-POLVIS SYSTEM AT SÈVRES À BOULOGNE BRIDGE

This experimental system was installed on the lateral storm weir situated on the right bank sewer at Sèvres bridge.

The basic principle is that floating matter should be retained in the sewer by means of a floating barrage, baffle, with a siphon wall; this baffle is laterally guided by two slide channels which adjust to the shape of the sewer section[2].

The equipment includes a grill which can be attached before discharging water in the Seine; the efficiency of the system during the test is thereby increased: baffle in place/baffle raised. The floating matter retained in the sewer is intercepted by the grills at the Avre station in Boulogne. The estimated duration of the test is about 6 months.

4. CONCLUSION

The objective of the study and research carried out until now in the sector of the Boulogne loop is to arrive at, in the near future, an optimal management of storm weirs leading to reduction of pollutants discharged in the Seine during the rainy season. The following steps have been taken towards achieving this goal:
— Linking all the information concerning storm weirs with the GAIA system.
— Constructing a simplified model of the sector of the Boulogne loop deduced from the detailed model, enabling definition of laws of functioning of the works in order to confirm the possibility of evolving a real time control system.

---

[1] See colour plate no. 6.
[2] See colour plate no. 3.

At the conclusion of these tests, the District can arrive at a real time system which will be helpful in making decisions for management of the sewerage network in the Boulogne loop; this is the goal of the European project S.P. 226.

DISCUSSION

Mr. Valiron: *Have you tried to or do you plan to **relate the values of depth and flow rate furnished by measurements on weirs,** to the impact generated by these overflows on the receiving environment?*

Mr. Soulier: *A project DHI-Safège on management in real time is in progress. It will cover the entire chain of decision-making, from pluviography and radar images up to the final decision. The study is to be completed in a year or two. The information gathered will assist us in making the final decision.*

Management of Interdepartmental Works of the Parisian Agglomeration

M. Le Scoarnec and Melle Genestine

1. INTRODUCTION

Around the year 1200, Philippe Auguste had the streets of Paris paved and a central channel for draining surface water built. Since that time the sewerage network has undergone development progressively.

Up to the end of the eighteenth century, there was no proper sewerage system per se and the waste water was allowed to flow to the Seine through a network of drains and 26 km of poorly maintained vaulted works.

Garbage piled up everywhere and was casually dumped in the Seine whose water was used for domestic purposes. A policy for general hygiene was only laid down after the tragic cholera epidemic in 1832.

In 1852 a law was promulgated stipulating that domestic and storm water of newly constructed buildings be poured into sewers (it was only in 1894 that all buildings were obligatorily required to install plumbing).

Until 1855, 110 km sewers in Paris were still discharging waste water in the Seine. Under the Belgrand project of 1856 sewers which could be inspected were planned, which would carry the dirty water downstream of Paris via the basins of Clichy (put into operation in 1875).

From 1895, water was diverted to the sewage farms at Gennevilliers (800 ha) and Achères (800 ha).

In 1910, 5000 ha of sewage farm under the aegis of Carrières-Triel and Méry-Pierrelaye began to receive the ensemble of Parisian effluents through a general drain 3 m in diameter.

Today, the Parisian network comprises 2100 km of sewers and the area of sewage farms is 2000 ha (farms at Gennevilliers are no longer in use).

Towards 1920, the sewage farms were found inadequate and biological treatment was initiated (already in the experimental stage since 1901). The proposed works were to be built at Achères and to be fed by 3.5 and 4 m diameter sewers.

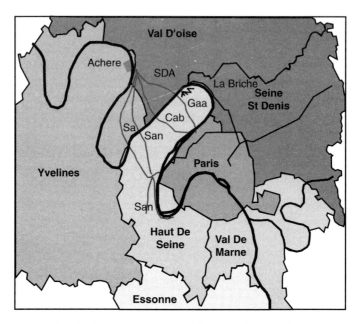

Fig. 1: *Network of circular sewers of 3.75 and 4 m diameter carrying effluents to the Achères works.*

This project was declared a public utility in 1935 but construction and operation began much later; several modifications were made in the original plan. It was finally completed on 4 May 1993 and began operation with water carried by the Sèvres-Achères sewer, branch of Saint-Cloud Nanterre.

SIAAP came into existence on 15 September 1970. Its territory is constituted by new districts created in 1968 (Paris, Hauts de Seine, Seine Saint Denis and Val de Marne) and by convention, 162 communities of 4 districts of the 'grande curronne' were associated. It serves 8.27 of the 10 million inhabitants of Ile de France. The management and construction of works are both under interdepartmental control; the districts would rather see them under departmental control.

New treatment works were constructed at Noisy Le Grand (1976), Valenton 1A (1987) and 1B (1992).

The programme 'Seine Propre' launched by Etat-Région extended over the period 1984-1994. One of its goals was to optimise utilisation of the interdepartmental network of transport of waste water set up by a system of telemanagement in real time of sewers—the SCORE system (System of co-ordination of organisation and regulation for utilisation of drains).

In 1992, when construction of the works planned under the General Programme of Sewerage (1935) was completed and those planned under the

270

programme 'Seine Propre' were nearing completion, SIAAP proposed a new 'Directive Plan for the Sewerage of the Parisian Agglomeration'. This plan includes the construction of three new treatment works, modernisation of the existing works, extension of the network of transport of waste water, treatment of surplus storm water and globalised and optimised management of feeding of the works.

By 2015, the capacity of treatment during the dry period is expected to reach 3.37 Mm^3 per day for the ensemble of the zone under the jurisdiction of SIAAP, whose production is approximately 3 Mm^3. A break-up of the expected capacity is given below:

Achères:	2.1	Mm^3
Valenton:	0.6	Mm^3
Les Grésillons:	0.3	Mm^3
Colombes:	0.24	Mm^3
La Morée:	0.07	Mm^3
Noisy le Grand:	0.06	Mm^3
Total	3.37	Mm^3

2. THE SCORE SYSTEM

2.1 MANAGEMENT OF EFFLUENTS

A study of flows carried out in the 1970s revealed that the installations for collection of waste water and the Achères treatment works are subject to the vagaries of water utilisation by the consumers. There are irregularities in the arrival of effluents, mainly at the Achères treatment works, where only a small quantity is received in the morning (underconsumption during the night) and an appreciably large quantity in the afternoon (overconsumption during the morning) (Fig. 2).

In view of the above observation, it would appear that regulation of flow is imperative to ensure as constant an inflow to the works as possible. One way to achieve this is to level the peaks by holding the large inflow until the next day; this is precisely what the SCORE system is trying to do.

As there is no possibility of constructing storage basins in the dense web of the Parisian agglomeration, regulation has to be accomplished by using the vacant portion of five drains for retention during high flow rates, by means of the 14 valves installed in the network.

The results obtained for the dry season during the four years of implementation of this plan are shown in Fig. 3. The curves show that for an identical mean flow rate over the day the troughs are almost completely filled. Given this dynamic

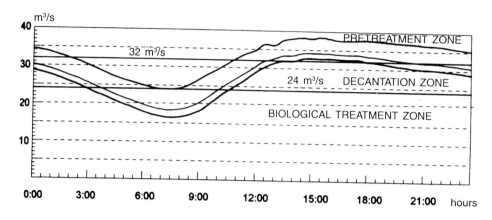

Fig. 2: *Curves showing irregularity of inflow of effluents based on data collected between 1990-1992. The lowermost curve refers to the dry season, the uppermost to the rainy period and the curve in-between the relationship between the two.*

In the dry season the minimum of 16 m³/s occurs around 7:30 h, the maximum of 33 m³/s around 16 h, with a mean for a typical day of 26 m³/s.

The straight horizontal lines show different levels of treatment at the works.

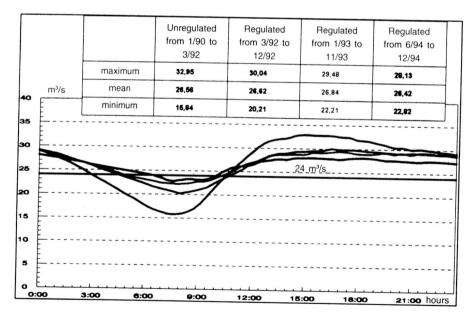

		Unregulated from 1/90 to 3/92	Regulated from 3/92 to 12/92	Regulated from 1/93 to 11/93	Regulated from 6/94 to 12/94
	maximum	32,95	30,04	29,48	28,13
	mean	26,56	26,62	26,84	26,42
	minimum	15,84	20,21	22,21	22,82

Fig. 3

272

utilisation of the capacity of the network per se, a larger quantity of effluents is now treated at Achères.

In 1991, the difference between the troughs and peaks was 17 m³/s; now (1994) it is no more than 5 m³/s.

Biological treatment (24 m³/s) has increased gradually over the four years:

    1991:   1,940,000 m³   (22.45 m³/s on average)
    1992:   2,026,000 m³   (23.45 m³/s on average)
    1993:   2,051,000 m³   (23.74 m³/s on average)
    1994:   2,064,000 m³   (23.88 m³/s on average)

The entire quantity of effluents goes through at least pretreatment or decantation.

2.2 EFFECT OF RAIN ON REGULATION

The storage capacity of drains is about 200,000 m³. In the phase of regulation, a large quantity of waste water is retained and the drains are of little help when the flow entering the network is very large.

During storm episodes the flow in the sewerage works increases due to the fact that the flow in a combined network is larger than in a separate network.

The networks already subjected to regulation do not, at present, have the requisite capacity for absorbing the increase due to storm-flow. However, using pluviometric radar imagery together with 'Calamar' it is possible to predict rainfall on the catchment area 1½ to 2 hours in advance. Taking advantage of this advance notice, the stored water is diverted to the treatment works, ignoring the process of regulation, and thus a capacity is created in the network to receive as much of the additional flow as possible. Evacuation is done at a rate determined by the intensity, velocity of displacement and direction of rain as well as by the maximum flow that can be diverted to the decantation and pretreatment units of the works.

This method enables maximum delay in overflow, even total suppression if the rain is not too persistent nor too intense.

2.3 IMPROVEMENT IN MANAGEMENT OF EFFLUENTS

Thus regulation of inflow to the treatment works of Achères by the SCORE system has been operational for the last three years and the results obtained are encouraging in view of the goal set at the time of design of the system, viz., stabilisation of inflow to the works throughout the day, especially in the dry season.

The performance of the system could be improved, however, were the following factors taken into account:

    — knowledge of the quantum of pollutants in the effluents in order to

273

distribute them in an optimal manner over the capacity of the works at different hours of the day;

— additional storage capacity for complete removal of the afternoon peak and morning trough;

— keeping in mind the possibilities and difficulties of the upstream networks which, by means of closely related systems, could ensure surveillance of their common and departmental networks;

— integration of works which could help in limiting the effects of rain in the sewerage complex;

— extension of the SCORE system to other equipment of SIAAP for overall management of effluents produced by the Parisian agglomeration.

The Directive plan of SIAAP, launched in 1992, takes care of many of the above factors and some are already in the course of implementation (works at Colombes, interlinking drains).

Other factors, such as limitation of overflow in the dry season, are still in the planning stage. As for generalisation of the SCORE system, it is still in the experimental stage. These two topics are discussed below.

3. SUPPRESSION OF OVERFLOW AT CLICHY

The study includes assessment of storage capacity and physicochemical treatment of surplus storm-water needed for suppressing the overflow at the Clichy works (Hauts de Seine).

Let us examine the behaviour of these installations during the past storm events as recorded by the SCORE system every half hour. The examination covers the period 16 March 1992 to 31 August 1994; the following intervals are excluded from examination, however: (1) period corresponding to non-functioning of drains of the Clichy-Achères branch of Argenteuil (CAA) and Clichy-Achères branch of Bezons (CAB); (2) period corresponding to pretreatment at the Achères works; during this interval the perturbed flow of effluents causes overflow.

During the period examined the amount of overflow at Clichy was 34,918,938 m^3, containing 10,476 tons of suspended matter (SS) and 3492 tons of BOD_5.

3.1 DATA ANALYSIS

The maximum rate of overflow during this period at Clichy was 59.9 m^3/s. This would require a treatment capacity of about 60 m^3/s, which would indeed be very costly.

274

To reduce the overall cost of the operation to an acceptable sum, the capacity of treatment can be minimised by increasing the storage capacity, which would even out the flow rate peaks.

3.2 METHOD ENVISAGED

For optimisation of the use of available means, the solution consists of combining treatment and storage.

The following method is recommended:
- Store water for one hour, the time required for necessary physicochemical treatment. In view of the small rate of flow during this hour, the storage capacity must be sufficiently large so that there is no direct overflow.
- Thereafter, allow the surplus water to flow into the storage works until a certain rate of impoundment is attained; the study covered different filling rates.
- When this rate is attained, two situations arise:
  — a larger excess of flow and consequently the effect of rain is also pronounced. The flow is treated in remediation basins up to their maximum capacity for treatment. The remaining part of the flow is allowed to fill the available part of storage;
  — the surplus flow is smaller and therefore the effects of rain also become less intense: during the subsequent period the situation may return to normal. The excess flow is then stored and once the maximum storage capacity is attained, the remaining flow is treated in the remediation basins in accordance with their maximum treatment capacity.
- When the situation normalises again, i.e., the drains can accept the entire quantity of water, it is expected that in the upstream network also the flow becomes normal (in the study, the time of restoration of normal flow was estimated as one hour). Thereafter the water present in the storage works is evacuated into the drains; in this study, the maximum rate of evacuation was 20 m³/s.

The advantage of this solution is that the water transport works are used for jettisoning part of the water which would normally overflow. Thus the overflow of pollution at Clichy, either directly or after treatment in the remediation basins, is significantly reduced.

The study covered 2880 possible cases (treatment capacity from 5 to 60 m³/s, with a step of 5 m³/s, combined with the possibility of storage from 0 to 600,000 m³, with a step of 50,000 m³ and associated with the rate of filling of 5% in 5%).

The solution which
— recommends suppression of every overflow observed during the period of study,
— and, after treatment, discharge in the Seine of minimal quantity of pollution (916 tons for the SS and 611 tons for the BOD$_5$ over the period considered),
— is the most economical (2100 MFrancs),
— involves a good rate of utility of the facilities (30.34% of storage capacity and 46.24% of treatment capacity),

leads to the following choice:

> **treatment capacity 25 m³/s**
>
> **storage capacity 450,000 m³/s**

The initial rate of impounding the storage work recommended for the defined use, is 15%.

It may be noted that the rate of occupation is defined as the barycentre of maximum storage, the balancing coefficient being proportional to the time for which the storage is utilised. Thus an equilibrium was found among:
— small successive overflows over a long period, leading to a slow but sometimes large storage;
— small isolated overflows leading to minimal storage;
— large isolated overflows, resulting in rapid and large storage;
— large successive overflows entraining a total mobilisation of storage capacity over a long period;
— etc.

3.3 STORM OF 18-19 JULY 1994

To confirm the soundness of the proposal, operation of the recommended facilities was tested over an intense storm event. During the rain of 18-19 July 1994 the inflow to the treatment works of Achères reached a maximal rate during the period of study: 63.64 m³/s as the average value over half an hour with a peak value exceeding 66 m³/s. The rate of overflow at Clichy also attained a maximum value of 59.91 m³/s, the total quantity of overflow being 1.3 million m³.

The Météo-France considered this event a rainfall with a 10-15-year return period. The performance of the chosen facilities was simulated upon conclusion of the previous study (storage: 450,000 m³, rate of treatment 25 m³/s) conducted during this event.

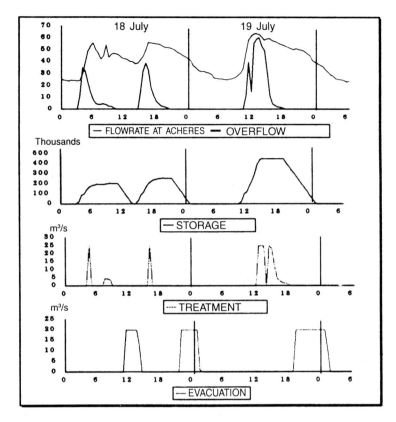

Fig. 4

Figure 4 thus shows:
— topmost curve, rate of flow reaching the Achères treatment works,
— second (from top) curve, overflow at Clichy,
— third curve, rate of simulating impoundment of a storage reservoir of 450,000 m³ capacity,
— fourth curve, rate of simulated treatment of a treatment works of 25 m³/s capacity,
— lowermost curve, evacuation towards the treatment works.

In this simulation, direct discharge into the Seine has totally disappeared while the overflow at Clichy during this rainfall was 1.3 million m³. The pollutants discharged to the environment would have contained 20 tons of SS and 13 tons of BOD_5 while the overflow on that day contained 390 tons of SS and 130 tons of BOD_5.

The remediation basins operated at full capacity during the period of large overflow; they were augmented by the storage works.

During the most intense storm episode (19 July, 11:00 to 18:00 h):

— the decontamination basins operated at full capacity for 2 hours (12:30 to 14:00 and 14:30 to 15:00 h),

— the storage capacity was utilised at 96.16% (from 14:30 to 15:00 h) for the highest treatment.

3.4 CHOICE OF FACILITIES

This simulation, during which the performance of the recommended facilities (storage 450,000 m^3, treatment 25 m^3/s) was studied, was found to be satisfactory.

However, the storage capacity was utilised at 96.16% for a rainfall with a 10-15-year return period and thus the situation was rather tight; it would be advisable to plan a somewhat higher margin of safety while making the final choice.

In every case studied the capacity of treatment of remediation basins was utilised to the maximum, for example during the phase of evacuation. It would therefore be futile to increase the treatment capacity.

Thus, to increase the safety margin it suffices to provide a larger storage capacity. In the case considered, with a capacity of 500,000 m^3 the percentage of storage utilisation is 86.5%.

The final solution recommended consists of the following choice:

---

treatment capacity 25 m^3/s

storage capacity 500,000 m^3

Due to the increase in storage capacity the recommended rate of initial filling changes from 15% to 30%.

4. OVERALL MANAGEMENT OF SEWERAGE

4.1 TELEMANAGEMENT OF WORKS

The four districts of the territory of SIAAP and SIAAP itself have in the past fifteen years studied and developed systems of telemanagement to assist them in utilisation of their own works as well as works entrusted to them.

The district of **Paris** studied the GAASPAR system (Automated management of the sewerage of Paris) for:

— its intramural network,
— hydraulic interface with works of SIAAP such as the basins of Clichy, North-eastern sewer and the South sewer first branch.

The district of **Hauts de Seine** developed the GAIA system (Management aided by informatics of sewerage) for:
— its own network,
— works entrusted by SIAAP,
— hydraulic interface with SIAAP sewers feeding the Achères works.

The district of **Seine Saint Denis** developed the NIAGARA system (New informatics for aid to automated management of sewerage network) for:
— its own network,
— works entrusted by SIAAP,
— hydraulic interface with treatment facilities of SIAAP such as Noisy le Grand works and Briche basins.

The district of **Val de Marne** developed the VALERIE system (Val de Marne, informed utilisation and regulation of effluents) for:
— its own network,
— works entrusted by SIAAP,
— hydraulic interface with the treatment facilities of SIAAP, such as the Valenton works.

As for **SIAAP** itself, it developed the SCORE system for managing, to begin with, feeding of the Achères treatment works.

The data of the new directive plan require extension of some system to take into account the works under construction as well as those yet to be constructed:
— NIAGARA for taking into account the Morée Sausset works,
— VALERIE for extension of the SIAAP network by adding the works VL and the 2nd section of Valenton,
— SCORE for creating a system of control of feeding the works at Colombes and Ferme des Grésillons,
— supervision of interconnecting sewers among various works of feeding treatment works.

All these 'real time' systems work independently for optimum management of their sector of influence.

4.2 EXCHANGE AMONG SYSTEMS OF MANAGEMENT

The systems of GTC can exchange information through an independent software called 'système partenaires'[1]. It is based on the exchange of files in accordance with a standard protocol of JBUS type; on arrival at each system it is translated into the GTC language.

---

[1] See colour plate no. 8.

Anyone can then obtain from a neighbouring GTC data in real time on the points that he does not possess and whose knowledge enables him to improve the system of his network[1].

Data files on variable periods can also be exchanged; this enables the acquisition of information stored in the 'partenaires GTC'. This system can interconnect up to 16 'clients'.

Each system is capable of managing its own sector and feeding the treatment works which are attached to it; on the other hand, it cannot act alone, independent of options selected by the other sectors. If it is ignorant of the upstream and downstream conditions—and at present such is the situation—its management cannot attain the desired efficiency.

The disparity among systems of telemanagement of networks imposes a necessary co-ordination through their goals as well as the facilities used. This gives rise to the concept of a federal organisation which can modify the decisions taken at the district level by independent systems since such decisions are based on only a partial vision of what is happening in the entire agglomeration.

The supervisory system COEUR (Co-ordination and optimisation of residual urban effluents) is expected to ensure the required synchronisation at the interdepartmental level. Its scope comprises:

— Proposing strategies of distribution of effluents among various treatment works: giving preference to the most modern works, removing the malfunctions of works, distributing the occasional discharge in the best manner.

— Preservation of the natural environment by managing in a global way such critical factors as not allowing the flow rate to exceed certain thresholds, filling or evacuating the basins and storage areas in accordance with well-defined criteria.

— Limitation of the adverse effects of storm overload which under certain conditions always result in more or less large overflows and are often devastating for fish. Météo-France's instruments of detection and prediction, integrated with the COEUR system enable anticipation of overload of networks and diversion towards the less loaded.

— Safety of the personnel responsible for the works in progress, limitation of flow rate and taking care of the water body.

This particularly reliable system will thus collect the main data on local networks, analyse them and propose a variety of solutions which are well suited to the local conditions of each telemanaged sector.

The proposed solutions with respect to the sewerage agglomeration will be recommendations and not orders and this central system should therefore be

---

[1]. See colour plate no. 9.

280

capable of modifying its strategy rapidly in case execution of directives proves impossible.

5. CONCLUSION

The management of transport of waste water produced in the territories of SIAAP is carried out by the five districts which are independent, each aided by a centralised technical system of management in accordance with its own principles, goals and methods.

The new data of the Directive Plan of the Sewerage of the Parisian Agglomeration will help this system grow so as to take into account the integration of new facilities and, in particular, improvement in the manner of utilisation.

To attain this improvement, some systems are equipped with software which enables optimisation of transport and distribution of effluents in its own zone of jurisdiction. Thus each sector is managed in accordance with its own 'best interests'.

In the coming years SIAAP will, to begin with, concentrate on improvement of the treatment of urban effluents through optimisation of the functioning of the Achères works by completing the works at Valenton and Noisy le Grand and establishing other centres of treatment such as La Morée, Colombes and the Ferme des Grésillons.

In addition, SIAAP is in some way the guarantor of the state of brooks, rivers and rivulets crossing the agglomeration and manages the floating barrages, large storm weirs and quality of water discharged from treatment works. Therefore it would be desirable that it propose installation of works of remediation capable of storing, treating and managing the surplus load due to rain and storms.

The agglomeration should not be managed sector by sector but holistically by the COEUR system, which is capable of determining the capacity of networks, has knowledge of the capabilities of the treatment and remediation works, and can anticipate the effects of storms and rain. In short, it is capable of better management of the treatment systems which SIAAP has decided to reinforce in its new Directive Plan of the Sewerage of the Parisian Agglomeration with the collaboration of its departmental partners.

DISCUSSION

Mr. Affholder: *No decision has yet been taken for implementing the final solution recommended at the conclusion of the study conducted by Melle Genestine.*

A study of the ensemble of the Parisian agglomeration is about to be launched; it will integrate the additional considerations regarding the the rainy season.

Mr. Valiron: *In Melle Genestine's presentation it was a question of evacuating the basins at the rate of 20 m³/s as soon as normal flow of the dry period was restored in the upstream networks. One could plan basin evacuation at a lower rate and utilise the facilities at Achères which are calibrated for flow rates lower than 70 m³/s.*

The storage works extend the period during which the facilities of treatment of effluents of the rainy season, situated downstream, need to be used.

This mode of development and management is preferable for two reasons: on the one hand, it does not involve the purchase of all the extensive equipment at Achères in one single instalment and, on the other, enables utilisation of the equipment over a longer period.

Mr. Fisher: *The installation of Clichy which was helpful for the study is unfortunately not the only one. Other points of large overflow have also to be equipped; the flow (at 45 m³/s) of surplus water of the rainy season admissible at Achères will soon be saturated.*

Mr. Affholder: *As a matter of fact, this study does not provide an answer to the question: 'What is the optimum rate of treatment at Achères' works?*

Mr. Gauvent: *When the rate of overflow at Clichy is 60 m³/s, the rate of inflow at Achères is 65 m³/s. So, there is a need to 'sponge' the flow arriving at Achères before allowing the quantity stored in the upstream basins to flow.*

Mr. Affholder: *What happens at the upstream of Clichy has not been fully taken into account in this study. All the weirs of Paris have yet to be equipped. Whatever is done in this regard will contribute to augmenting the flow rate or, in any case, the quantity of effluents arriving at Clichy. The network's capacity for storage and transport will then be maximally utilised and that part of the overflow from the present weirs, which is not allowed into Paris, will converge to the Clichy site.*

The simulations made use of the hydrographs corresponding to a mode of management of the upstream catchment area, which differs from the mode of management to be used in future. Thus for the same storm event a larger quantity of effluents is now arriving at the Clichy site.

Optimisation of the ensemble of the facilities thus requires simulations which are more global and integrate the capacities of the ensemble for treatment and rules of management applied throughout the network.

Treatment Plant of Achères: A New Directive Plan on the 2001 Horizon

M. Gousailles, SIAAP

INTRODUCTION

SIAAP (Interdepartmental syndicate for sewerage of the Parisian agglomeration) is a public organisation responsible for the transport and treatment of the waste water of the Parisian agglomeration. It serves an area of 1900 km², supporting 8.2 million inhabitants. There are several installations, the largest being the well-known treatment device of Achères.

This plant consists of several sections whose nominal çapacity today is 2.1 million m³ per day. It operates on the principle of an activated sludge process, is fed by a combined network and removes organic pollution of waste water of the dry season. It was not designed for removal of nitrogen or phosphorus, however, nor for handling the surplus water during the rainy season.

Consequently, water quality in the Seine, downstream of the discharge, is altered in the severe low water situation and during intense rainfall.

In order to increase the impact of the treatment works on restoration of the natural environment, SIAAP has formulated a new plan for development of the installation which ought to augment its performance during both the dry and rainy seasons.

1. PRESENT SITUATION

1.1 TREATMENT WORKS OF ACHÈRES

After a common preliminary treatment, the four sections of the plant carry out primary sedimentation followed by biological treatment by activated sludge. The biological treatment differs slightly from one section to another in terms of the period of hydraulic retention time (1.5 to 3 h) and mode of feeding the settled water, but in every section the load is quite high—between 0.6 and 1.2 kg BOD$_5$ per kg of VSS per day.

The principal activities of the installations are given in Fig. 1 and the breakdown of raw water and treated water during the dry season in Fig. 2. During the rainy‑ period the surplus water is pretreated and a small part settled.

	AI	AII	AIII	AIV
Time of hydraulic retention	3 h	3 h	1.6 h	2 h
Overflow rate on secondary clarifier, m/h	1.70	1.90	1.25	1.60

Fig. 1

mg/l	SS	BOD$_5$	COD	TKN
Raw water	250	200	400	43
Treated water	90	30	90	37

Fig. 2

1.2 IMPACT ON NATURAL ENVIRONMENT

In the Seine, which receives effluents from the Achères works, the flow is meagre; the usual value at Paris for the quiquennial low water over 30 consecutive days is 55 m³/s. The mean waste-water discharge of the dry season is 25 m³/s and induces, in these conditions, a significant degradation of the water quality. Figure 3 shows the profile along the Seine of major parameters of quality obtained by mathematical modelling.

A large deficit of dissolved oxygen has been observed at the downstream of Achères. This deficit is due to removal of oxygen caused by natural decomposition which, at the first stage, is related to BOD$_5$ and at the second stage, ammoniacal nitrogen. It has further been observed that nitrification begins quite late, implying that a relatively high concentration of ammoniacal nitrogen subsists in the lower Seine, which could by itself lower the concentration of dissolved oxygen, a risk of toxicity to fish.

During the rainy season, a discontinuous situation, modelling is more complicated. Figure 4 shows the measured value of oxygen in the Seine as recorded by a station situated about 30 km downstream of Achères. The Figure

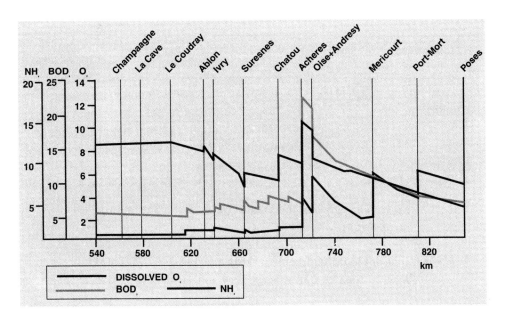

*Fig. 3: Profile along the Seine
(as of 1995, at 55 m³/s).*

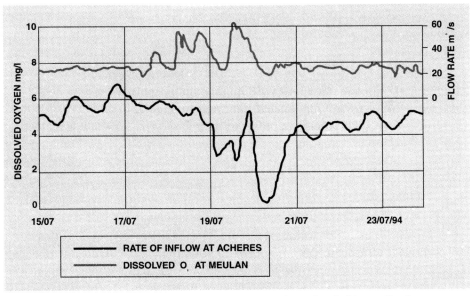

*Fig. 4: Dissolved oxygen in the Seine after the Achères discharge
during the rainy season.*

also shows the flow rate from the treatment works. The impact of overflow of surplus water discharged is even more marked at the measurement station, in spite of the confluence with Oise and the effect of reoxygenation of the barrage of Andrésy. The deficit of dissolved oxygen, compared to the dry season, is attributable to organic pollution, which is the only quantity that increases during the rainy period. The long duration of the impact means that much of the discharged pollution is in dissolved, particulate, non-decantable or very slowly decantable form.

Thus it would appear globally that the deficit in the river is primarily of oxygen resulting from the discharge of oxidisable organic or nitrogenous matter. Since both organic and nitrogenous pollution result in an oxygen deficit, a new parameter has been proposed—the GOD (global oxygen demand)—in order to readily quantify the level of a pollution. This parameter is only an indicative one because it has different spatiotemporal implications and is defined as follows:

$$GOD = BOD_5 + 4.5 \, (TKN - 2)$$

In fact, experience has shown that there subsists in water about 2 mg/l of non-decomposable organic nitrogen, resistant to ammonification.

During the dry season, counting on an efficiency of treatment of 85% for BOD_5 and 15% for Kjeldhal nitrogen, the efficiency in terms of GOD is of the order of 53% for a daily flux of 800 tons.

During the rainy season, assuming that the quantum of nitrogen remains unchanged, the efficiency of GOD on a typical day falls to 34% for a flux of about 1100 tons.

2. PROBLEMS OF THE DRY SEASON

Until the last few years the main preoccupation of the authorities was protection of the natural environment during the dry period; the problem of surplus water during the rainy season did not constitute a priority. This prompted SIAAP towards the end of the 1980s to launch an overall evaluation to understand the technical and financial conditions for improving the quality of water treated by the Achères works.

This evaluation covered only the AI and AII sections in which contamination was lowest and the conclusion was that it would be more economical to opt for tertiary treatment by fixed cultures than to extend the treatment with free cultures. On the one hand, this conclusion was supported by the existing networks which do not enable economic integration of denitrification and, on the other, extension of this exercise to AIII and AIV sections in which the waste is stronger, proved to be financially even worse with respect to the activated sludge process.

Since evaluation conclusively favoured fixed cultures, three procedures of tertiary treatment whose performance had been preliminarily tested at CRITER, were selected for tests at the industrial scale.

Three industrial prototypes of the procedures, Biofor, Biostyr and Sessil, were constructed. Although denitrification was not the priority goal of SIAAP, it was stipulated that the possibility of an ulterior denitrification be kept in view.

Two prototypes of immersed filters, Biofor (Degrémont) and Biostyr (OTV), are currently being tested at Achères while the prototype using a bacterial bed Sessil, followed by a lamellar decanter (Stéreau), is being tested on the site of the Noisy le Grand treatment works. The latter was installed only recently and therefore it would be premature to include the results of this test.

The main characteristics of the prototypes are briefly depicted in Figs. 5, 6 and 7.

Without detailing the results currently available, it would nonetheless appear that these devices have considerable flexibility of operation. Knowing the quality of the water to be treated, it is thus possible to obtain a desired quality of treated water by applying the optimum load. Figure 8 shows, for one of the biofilters, the pattern of the curve of quantity of pollution removed as a function of the initial concentration. It is possible to construct from this curve a chart showing the relationship, in the given conditions, between the velocity of water and the concentrations at the inlet and outlet of the biofilter (Fig. 9).

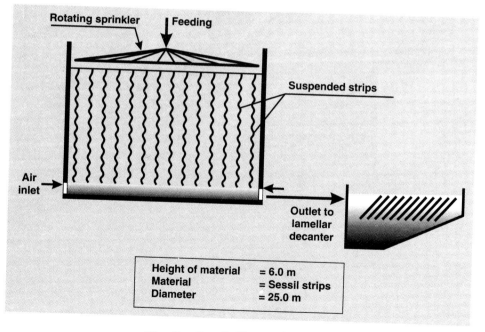

Fig. 5: Sessil (Stéreau) prototype.

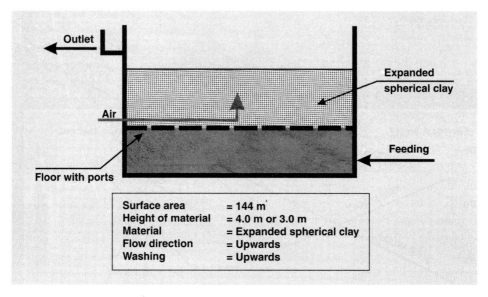

Fig. 6: *Biofor (Degrémont) prototype.*

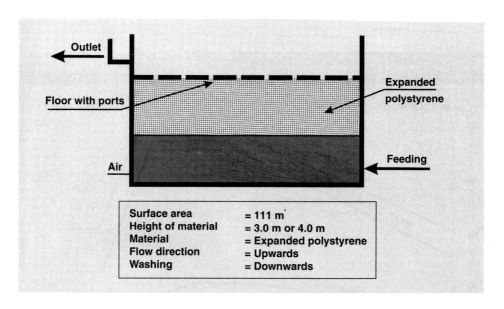

Fig. 7: *Biostyr (OTV) prototype.*

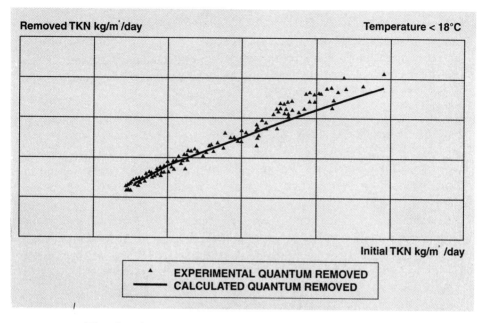

Fig. 8: *Curve expressing the quantity of TKN removed as a function of its initial concentration.*

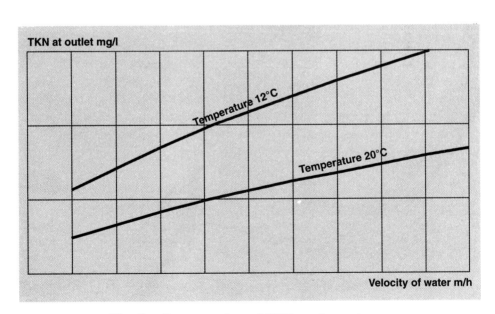

Fig. 9: *Concentration of TKN at the outlet as a function of the velocity of water.*

It was found that the goal of 5 mg/l of TKN in the treated water was tenable, with a corresponding BOD_5 of about 10 mg/l. Thus an efficiency of the order of 92% removal of GOD could be expected.

The experiments on the three prototypes were conducted to gather knowledge which would enable SIAAP to make the best technoeconomical choice when required.

3. PROBLEMS OF THE RAINY SEASON

It was only in the early 1990s that the need for treatment of the pollution of the rainy period was recognised, especially in the light of the European directive.

In the Achères works, the resultant small residence time and the large overflow velocities on the clarifiers prevent overloading with respect to flow rate of the rainy season. As a matter of fact, the rate of flow into the treatment works tends to increase rapidly up to 70 m^3/s while the mean flow rate of the dry period is 25 m^3/s.

In these conditions, the excess flow rate necessitates the employment of means which can respond instantaneously to a large excessive flow rate. At the conclusion of a new evaluation, SIAAP made the choice of installation of lamellar decantation after flocculation and coagulation of effluents.

The purpose of an installation for this type of treatment is removal of suspended matter with 80% efficiency. It is also expected to eliminate about 60% BOD_5 from the physicochemically treated water.

Figure 10 shows a flow diagram of the procedure chosen by SIAAP. After fine sieving, water enters a lamellar decanter (procedure Actiflo OTV). The decanter has zones of conditioning ($FeCl_3$, polymer) and incorporation of sand (diameter: 135 μm). The particles of sludge weighted by the sand get deposited at the bottom of the decanter. A hydrocyclone carrries out dirt-sand separation. The dirt is evacuated to a thickener and the washed sand is recycled. The equipment is designed for the best performance at a surface overflow velocity of 90 m/h.

4. NEW DIRECTIVE PLAN FOR THE ACHÈRES WORKS

4.1 DESIGN

For the sake of simplicity, the consideration by SIAAP of the problems of the dry and rainy seasons were presented separately and chronologically. In fact, improvement of knowledge gained through tests on the prototypes and a firm desire to utilise to the maximum the capacity of treatment works, prompted

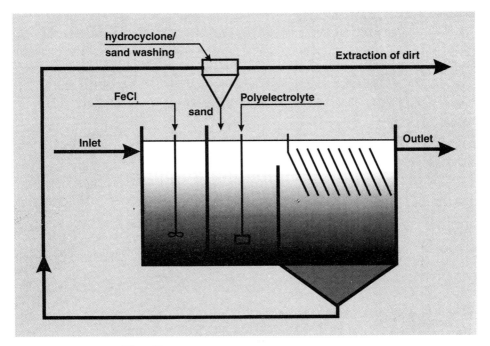

Fig. 10: *Actiflo: efficiency for SS = 80%.*

SIAAP to formulate a new directive plan for the treatment works at Achères, by integrating the problems of the dry and rainy periods.

Tests on the prototypes installed at Achères, the biofilters, have already shown that this type of arrangement is quite capable of handling the peak flow rates and the quantum of organic pollution without risk of malfunctioning.

To obtain the results desired during the dry season, as already described, it is necessary that the flow velocity in the biofilter be significantly lower than the hydraulic limit of the process. Thus the biofilters have a very large hydraulic capacity which is immensely useful during the rainy season when the flow rate increases by a factor of about 3.

The new directive for the Achères works thus aims to couple the physicochemical installations for the rainy period and the devices of tertiary nitrification and refinement during the dry season. Such coupling has numerous advantages, especially in the rainy season.

It has been seen that physicochemical treatment enables a good reduction of SS but is less effective for BOD_5 during the rainy season; in terms of GOD during a typical storm event, the efficiency would be only of the order of 40%. In view of the observations made on the Seine, this efficiency appears inadequate for preventing large deficits of dissolved oxygen. On the other hand, test results show that the effluents of physicochemical treatment can be admitted to the biofiltration process after mixing with water treated by activated sludge device. The overall efficiency of GOD would thereby be highly increased due to reduction of BOD_5 and partial nitrification of ammonia. Further, the suspended matter would be removed by filtration. The behaviour of a biofilter in a simulated storm is shown in Figs. 11 and 12. BOD_5 and TKN reduce significantly, thereby demonstrating the efficiency of biofilters in these conditions.

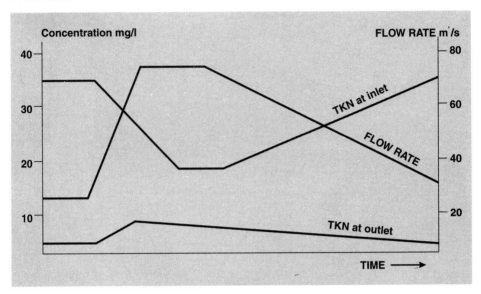

Fig. 11: *Biofilter: simulation of TKN during a storm.*

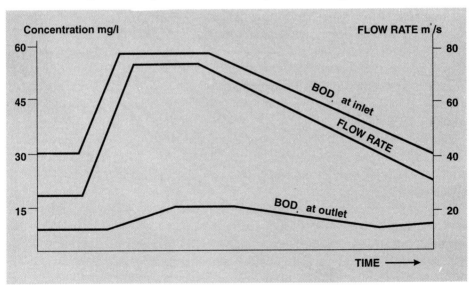

Fig. 12: *Biofilter: simulation of BOD_5 during a storm.*

294

This possibility of obtaining high efficiency of treatment during the rainy season amply justifies treatment of surplus water in the Achères plant rather than on other sites upstream. It is also possible to plan phosphate removal from water of the dry season by the physicochemical method, to reduce the flux of nutrients flowing into the North Sea.

4.2 CONSTRUCTION, GOALS

Construction of the first section of treatment of surplus water of the rainy season at the rate of 22.5 m³/s, has already been initiated by SIAAP. Construction of the second section will be undertaken as soon as performance results are in[1]. It will then be possible to extend the capacity to 45 m³/s.

As already mentioned, construction of installations of tertiary nitrification will also be undertaken shortly thereafter.

These facilities will enhance the performance of the Achères treatment works.

The developments planned for dry and rainy periods under summer conditions are shown in Figs. 13 to 15. As an assistance to the reader, the daily flux of GOD at the inlet and outlet of these installations is also shown for three different situations.

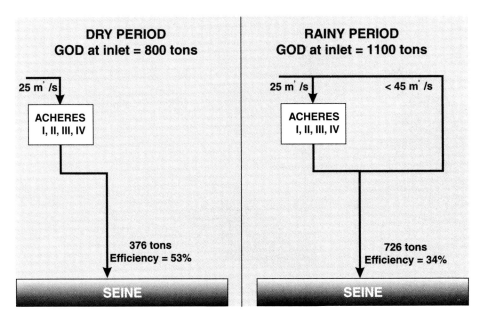

Fig. 13: Situation in 1995: present state of installations.

---

[1] This report will be part of a study on discharge in the agglomeration; results are expected by mid-1997.

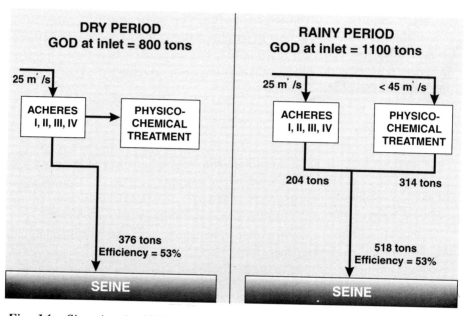

Fig. 14: *Situation in 1998: physicochemical treatment of surplus water during the rainy season at the rate of 45 m³/s.*

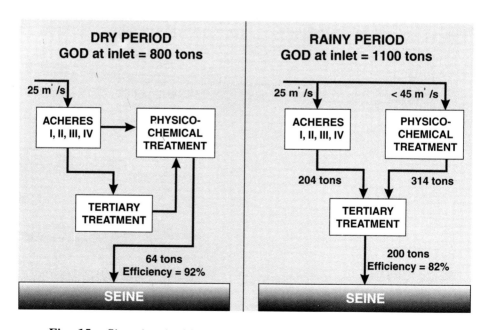

Fig. 15: *Situation in 2001: tertiary treatment implemented.*

CONCLUSION

The new directive plan for the treatment works at Achères described above is expected to be operational by 2001. Construction takes into account the particular situation of the river whose most constraining parameter at present is the quantum of dissolved oxygen. The goals of reduction of the global oxygen demand are based on results obtained from simulations carried out from data furnished by tests on a full scale. They are thus entirely realistic and take into account the precautions taken during simulation; it is expected that these goals will be surpassed.

The expected large reduction in pollution by this programme is due to the employment of biological means which can be overloaded during the rainy period. Maximum attention has been paid to treatment of water during the rainy season. In fact, only physicochemical treatment, envisaged at other sites, cannot yield comparable results.

Furthermore, should the plant undertake the process of reduction of flux of nutrients, the procedure of phosphate removal from the water of the dry season will also achieve denitrification.

As for finances, the cost of the entire operation has been estimated at 4000 million francs, which certainly is a very good cost-efficiency ratio.

Strategy for Overall Management of Sewerage in Seine Saint Denis

M. Breuil and C. Cogez
(Directorate of water and sewerage)

PREAMBLE

Since 1992 a number of laws and bylaws have been promulgated and circulars issued which precise the responsibilities and powers of the personnel of public services with respect to protection and management of the natural environment. These promulgations are concerned with sewerage and are based on certain basic concepts (urban residual water, storm weir, size of catchment area) which help in defining personnel obligations.

It is important to understand the gap between this theoretical approach and the reality as well as to give serious thought to various types of sewerage systems.

Using the catchment area 'Vieille Mer-Morée' to illustrate the situation, strategies of global management implemented in Seine Saint Denis are described. Starting with maintenance of quality and the final destination of effluents, the description proceeds with strategies for construction and day-to-day management of the works.

1. COMPLEXITY OF NETWORK CONFIGURATION: EXAMPLE OF THE VIELLE MER-MORÉE CATCHMENT AREA

1.1 Types of effluents (overflow of combined system or storm-water?)

The map in Fig. 1 shows four zones divided on the basis of types of effluents:

(1) In the downstream part, the Vieille Mer crosses a combined zone at Saint Denis. It deals with the overflow of the combined network at various points.

(2) Slightly upstream of the Vieille Mer lies the outlet of the separate catchment area of Saint Denis and Stains.

(3) The inflow from Molette contains the water of overflow of a combined catchment area, well-decanted in the retention basin of Molette.

(4) The complementary inflow from two districts, Croult and Morée, drains into the vast separate type catchment areas which have a large capacity (nearly 2 million m³). The basin of Brouillards enables decantation of the flow. The discharge of STEP of Bonneuil and Morée can be seen on the map. It may also be noted that in view of its malfunctions, the separate network receives at several points the concentrated waste water during the rainy period.

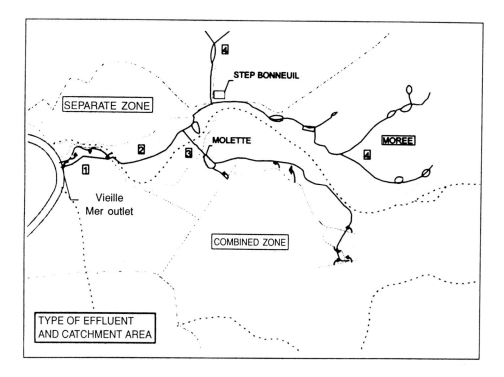

Fig. 1

1.2 CONCEPT OF CATCHMENT AREA

It is clear from Fig. 1 that accurate dimensioning of the catchment area is difficult:

— should one include the overflow of the combined catchment area which relates to other outlets?

— should one include for the ERU the separate zones for which, during the rainy season, a significant part of domestic effluents diluted by storm-water flows towards the Vieille Mer?

1.3 CONCEPT OF WEIR

This concept often appears in legislative promulgations which, depending on the quantity of discharge, list stipulations for communities. These works are strategic for the manager when deciding whether to direct the effluents towards the natural environment or to a treatment device.

Figure 2 shows the location of key points for Vieille Mer:

1) two points of discharge in the river, ultimate outlet of the supply basin during the rainy period;
2) a number of combined storm weirs on the course of the Vieille Mer (6 on Saint Denis, 10 upstream of Molette);
3) points of management of watercourses between the natural environment and treatment works, such as 'pumping at Molette';
4) points of overflow, due to storm, of waste water towards storm water, e.g. at Balagny (where the treatment works of Morée will be erected);
5) the plant of Bonneuil connected with the network for discharge of the dry period and weirs of the rainy season. Treatment is designed in accordance with the norms of discharge to the natural environment.

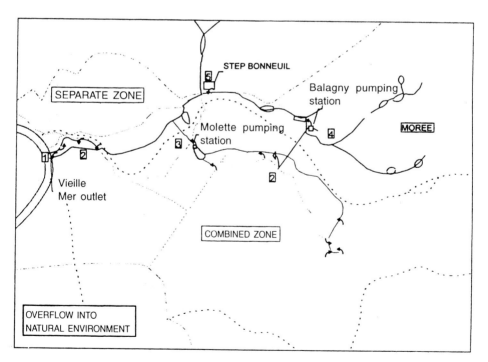

Fig. 2.

301

This concept is far removed from ground reality. The goal of the manager is to limit the discharge to the natural environment. The networks transit different types of effluents at different times. Real situations are illustrated in Fig. 3.

For example, the Bondy-Blanc-Monsil drain (1) designed from the beginning to carry combined overflow, uses the network (2) to transit waste water of the dry period via the pumping station at Molette, then (3) returns to the normal circuit.

On the other hand, the pumping station at Molette, built for requirements during the dry period, is used during the rainy period to divert the water evacuated from the Molette basin (the most polluted) to the treatment works.

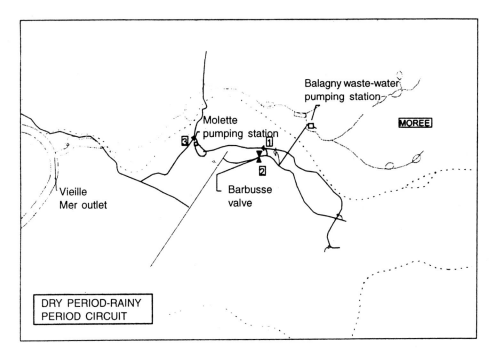

Fig. 3.

2. PERMANENT UPKEEP OF THE DESTINATION OF EFFLUENTS

In view of the difficulty of characterising by means of hydrological parameters the quality of effluents which transit in the networks, continuous measurement is an indispensable tool of knowledge

The total amount of overflow during the dry period in the course of the first six months of 1994 was 274,610 m³, of which 145,690 m³ was contributed by programmed operations. Again, of the 145,690 m³, the largest part (76,300 m³) corresponded to discharge into the Seine, from the North sewer for interception of sewers D11 and DD11 in February and March; during this period the North. East drain did not allow diversion of the flow from the city of Paris.

The total amount of overflow during the rainy period in the first six months of 1994 was 396,480 m³. The lower value, as compared to the first six months of 1993, is due to the absence of a significant storm event during the first six months of 1994, while during the second half of 1993 there was particularly heavy rainfall: 32 events, of which 6 corresponded to rainfall higher than 10 mm.

Distribution according to source of the amount of overflow during the dry period for the first half of 1994

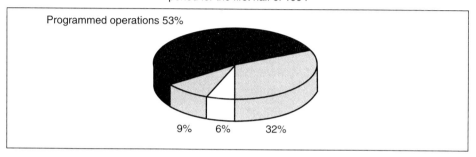

Interannual comparison of trimestrial overflow of the dry period

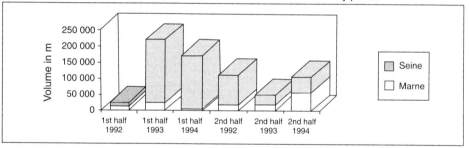

Interannual comparison of trimestrial overflow of the rainy period

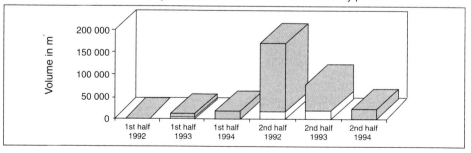

Fig. 4: *Example of overflow distribution in the telesupervised network: synthesis for the first half of 1994.*

The Seine Saint Denis region poses a very limited problem for the natural environment as there are very few points of significant discharge.

At the upstream of these points of discharge, in the centre of the district, there are 82 points of connection between the dry period circuit and the outlets of the rainy period. These storm weirs or safety weirs are considered important because they help in detection of abnormal diversion.

Lastly, there are 300 nodal points (valves, pumps, barrages with shutters) on the network of the dry period which make diversion of the effluents to one of the drains possible, depending on the requirements for maintenance of the works.

This ensemble is complex and it is presently not possible to control the information in real time. But it is possible to make monthly (or over several months) assessment of the amount transported or diverted through the facilities available for management of water. Assessments have been made for 22 of the 82 connecting weirs, most of them towards the downstream, including 7 towards the natural environment. Thus a significant value for the amount of overflow and its source can be obtained (see Fig. 4).

On the other hand, exchange of data with the SCORE system of SIAAP helps today in ascertaining the amount diverted towards the Saint Denis Achères drain and soon will show diversion towards the Noisy le Grand treatment works.

The picture of discharge in the river is not complete, however. Even before the new legislative measures regarding discharge in the natural environment were promulgated, the District had initiated a programme for installing 23 stations of surveillance of discharge in the natural environment (SSRMN) during the period 1993 to 1997.

These stations will enable:

— knowledge in real time of the flow rate and the load of pollutants flowing into the Seine and Marne with the possibility of an alarm in case of abnormal discharge;
— overall assessment of discharge soon after a storm event;
— assessment over longer periods.

There are thus a number of goals:

— management in real time of works with quantitative and qualitative objectives;
— assessment and knowledge of phenomena on the significant events, study of projects;
— long-term analysis of factors influencing the network;
— basic information for statistical study and research on impacts.

The series of measurements considered necessary, involve a number of studies and innovations:

— Measurement of water quality in the network is approximated by direct measurement of turbidity. This procedure necessitates further improve-

304

ments for overcoming the problem of probe clogging.

— On-site acquisition of water quality data in a specific manner so as to improve the filtering of aberrant measurements and validation of data.

— Lastly, development of software for monitoring and utilisation of measurements made at these stations (VARENE and IRENE) in order to obtain the information required for the assessment mentioned above.

3. STRATEGY FOR MANAGEMENT OF QUALITY OF DISCHARGE

The overall goal of the District is reduction of impacts of discharge on the natural environment through a progressive management leading to enhancement of efficiency and economic solutions. Effort has also be be made to implement a mode of management that is flexible and considers future requirements.

Thus it should be ensured during the dry period that there is no discharge of waste water during the critical period of May to October. This implies planning and programming implementation of shut-downs in co-ordination with the community, SIAAP and the other Districts.

As seen in the example of Vieille Mer, all available means are adopted to ensure satisfactory transport of flow, irrespective of the source type (waste water, storm-water, both).

For the rainy period, the departmental directive plan envisages different strategies which take due account of the characteristics which differ widely from one catchment area to another (Fig. 5).

3.1 Morée Sausset zone

This is a vast separate catchment area with an outlet of small capacity (10 m³/s or 4 l/s per impermeable hectare). It has a very large storage capacity (which will double in future, either through reinforcement of the network or extra storage to be constructed). It is located at a large distance (30 km) from the treatment works.

The present storage capacity is quite sufficient to ensure extensive decantation upstream (followed by complementary treatment of the flow downstream with the basin of Brouillards of 80,000 m³ capacity).

To bring the decontaminated water to the Seine 4 km away, a drain 'Garges Epinary' may be used after some alterations to achieve better separation of the treated and untreated works.

This arrangement would also make it possible to bring 'an outlet towards the natural environment' around Val d'Oise and thus enable the Syndicate of Croult

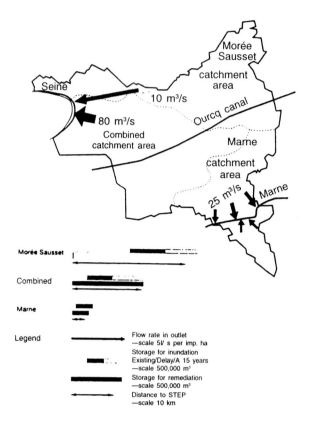

Fig. 5.

(SIAH) to independently pursue its objective of remediation of the dry period flow (works at Bonneuil) and the rainy period waters (1 million m³ of storage-decantation).

3.2 COMBINED ZONE OF EX-DISTRICT OF THE SEINE

A vast network of works which have facilities for inspection, are intermeshed and include storm drains of large capacity. There are fewer retention basins and the rate of discharge flow is relatively high (equivalent to 18 l/s/ha).

The treatment works are at a distance of about 20 km from this zone and therefore it is presently futile to plan diversion of storm-water flow towards Achères, which amounts to about 80 m³/s; the capacity of Saint Denis Achères is 12 m³/s. Instead, the flow from various parts is intercepted in the neighbourhood of banks and regrouped to be transported to devices with large capacity. Treatment of such huge quantities at an urban site would perhaps imply the use of more compact techniques (flocculation, lamellar decantation etc.).

306

A plan of remediation is under formulation for transporting effluents of the rainy period by means of large drains to devices for treatment of dissolved pollution. In this case, storage will enable lamination of the flow at the least cost. This will require strategic guidelines for management in regulating the flow into the drains without evoking significant hydraulic risks.

The basin of Plaine, under construction at the Grand Stadium site, illustrates this situation.

3.3 MARNE ZONE

This zone comprises multiple catchment areas for high floods.

The present storage capacity is small but will be increased within the next few decades to limit the risk of inundation. The treatment works of Noisy le Grand are located on the catchment area itself and therefore the distance is short.

In this case, the solution chosen consists of providing a treatment at STEP of about 30% of the decennial flow.

A storm basin on each intercepted sewer enables temporary storage of surplus flow before it is diverted to the treatment works.

To complete the discussion of strategic aspects of the departmental action, the reader's attention is drawn to the curve (Fig. 6) relating the volume required upstream of the network to control the flow rate (combat inundations), with the volume required downstream for remediation (consisting of simple decantation for two hours).

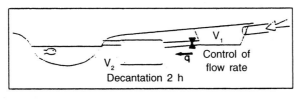

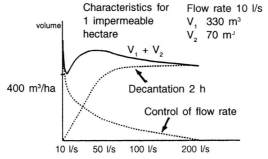

Fig. 6.

The adjustable parameter is the rate of flow which transits between the two basins (V_1: upstream storage basin, V_2: downstream basin of decantation before discharge).

The curve $V_1 + V_2$ is an indicator of the investment to be made. It passes through a minimum (about 400 m³ per impermeable hectare). About 80% of this volume is to be provided for storage upstream and 20% for remediation downstream. The difference would be even more marked if one includes the cost of sewers and plans for a more expensive remediation (Fig. 6).

It can thus be seen that in view of the goals of remediation during the rainy period, large storage capacity has to be provided. The department has to make a choice between constructing basins to take care of the hydraulic inadequacy of the present network and compensating the effects of new urbanisation. This equipment will form an integral part of the plans for remediation of the rainy period discharge from the Parisian region.

DISCUSSION

Mr. Valiron: *With what instrumentation does the Seine Saint Denis plan to equip the complementary measurement points (hydraulic and pollution flux)?*
A study covering the ensemble of the Parisian agglomeration is about to be initiated; it will include special consideration of the rainy period.

Mr. Breuil: *The Seine Saint Denis has chosen **simple equipment** for this purpose. Some sites have already been equipped with devices for measurement of depth and flow rate and have been integrated in the system of automated management. Their maintenance has been more or less standardised. The rule adopted consists of **preserving the existing equipment and supplementing it wherever considered necessary.***
The main problem encountered in measuring pollution flux is probe clogging. Should the problem prove unresolvable, Seine Saint Denis may have to think in terms of more expensive equipment, but this could mean a reduction in number of measurement sites.

Mr. Tabuchi: *What is represented by the 70 m³ per impermeable hectare that you propose to treat?*

Mr. Breuil: *The present approach is theoretical. We chose to make an assessment of the investment to be made from upstream to downstream when a constant flow is treated by decantation over 2 hours.*

If a more extensive treatment is planned at the downstream—which will undoubtedly involve higher expenditure—this will definitely have a marked effect on financial planning. A larger storage will thus need to be provided upstream to minimise the rate of flow downstream, in order to reduce the overall cost. A study may be made of the feasibility of **modifying the destination of effluents depending on their concentration.** *In fact, if more storage is provided to regulate the flow downstream—to around 10 l/s/ha—the residence time of effluents in the retention basin for decantation would have to be extended.*

Perhaps it would be better to manage the effluents in accordance with their destination—as is currently being done at La Molette—and bring more concentrated effluents to the treatment works.

Mr. Goussailles: *You add the volume of decantation and volume of storage, which is also the volume of decantation.* **What is missing is the respective costs of construction of one m³ of storage and one m³/s of capacity of water treatment.**

Mr. Fisher: *I would like to know the* **goal to be pursued and the performance level expected with the planned equipment.**

Mr. Breuil: *In the example cited, I assumed a certain type of remediation downstream. If another type of installation is desired downstream, one has simply to replace the equipment used in my example with another type and ascertain the rate of flow from upstream to downstream, which would optimise the overall cost. The purpose of my simple theoretical example was to give an order of magnitude of the overall cost.*

Mr. Affholder: *As for teatment, there is no unique solution for handling each type of discharge.* **One must think in terms of an 'ensemble of discharge' and also an 'ensemble of collection-treatment'** *in order to realise an overall view. Be it a matter of decantion for 2 hours or biological treatment, generalisation on the basis of a particular type of treatment is incorrect.*

Mr. Breuil: *That was not the intent of my presentation. To ascertain which type of decontamination will be required, one has to start with the natural environment and* **proceed from downstream to upstream.**

The order of magnitude of storage to be planned simply enables us to estimate the amount of work that needs to be done.

General Summary and Conclusions

At the end of the presentations and discussion, Mr. **Affholder,** chairman of the afternoon session, invited comments from the audience.

Mr. **Valiron** posed the problem of **visual pollution.** He suggested that it be given serious thought by a team led by AGHTM. He further added these three comments.

— **Hydraulic measures on weirs.** Unresolved problems require further work. Of course, it will be possible in the near future to conduct reliable measurements on the ensemble of the network.

— **Measurement of pollution flux** seems to be more problematic. The systems currrently available are rather complicated. Perhaps a changeover from the present dynamic hydraulic management to a more sophisticated dynamic management will become possible in the next few years.

— Furthermore, it is certainly worth thinking of a system with broader scope than that of **'Partenaire'** to enable concerted decisions to be taken very soon with respect to choice of modes of management applied to the entire network during dry and rainy seasons. Such a system should have maximum autonomy for the present decision-makers, i.e., the various managers. In some cases this system could formulate regulations regarding the life style and common activities of inhabitants.

Mr. **Affholder** also felt that efforts should be continued to improve the design, functioning and mode of management of storm weirs. He further mentioned that various managers of the networks of the Parisian agglomeration ordered, sooner or later, the same type of equipment and used comparable models; thus one could hope for rapid conclusions.

He felt, on the other hand, **that interception of buoyant matter** continues to pose a problem. The experiments in progress in Hauts de Seine will without doubt yield useful results. But definitely, there must be other solutions. Efforts towards experience must be generated if the problem is to be resolved.

In reply to Mr. Valiron's reference to experiments conducted abroad, he mentioned that there are many systems of screening, but such equipment is not readily usable on storm weirs.

With respect to tools of knowledge and management, he pointed out that there is a **central system, different from the informatic system, namely, the receiving environment,** which takes in practically the entire discharge. It is this integrating system which will assist in the changeover from a phase of exchange of information in which everyone manages his own network, to a system which integrates the essentials of the centralised system, leading to optimisation of utilisation of the capacity for treatment, storage and transport.

This stage has yet to be reached because the capacity of treatment downstream is inadequate; but one must prepare for the future. Various relevant services will then ensure a **true co-ordination of action** with the choice of reconciling the goals set by the receiving environment as well as the general and local goals.

In response to observations made by Mr. Valiron concerning management of storage and utilisation of capacities of treatment, Mr. **Gousailles** pointed out that there are two solutions to mitigating the nuisance generated by storm weirs: increase in the capacity of transfer, undoubtedly complex and provision of compensatory measures.

On the other hand, the solution consisting of increasing storage capacity becomes difficult to apply where this is a large difference between instantaneous flow to a point and the capacity of evacuation to a destination, say a treatment plant.

He felt that if arbitrary management of a network leads to a reduction in capacity of evacuation or transfer of a work, this would result in an increase in risk of overflow and enhancement of the requirements of storage or treatment.

Mr. Valiron pointed out that to face the risks of increase in overland flow as a consequence of new urbanisation, it should be stipulated that the **builders monitor the quantity of discharge in the outlet of the land after construction.**

In response to Mr. **Gousaille's** remark, he said he was aware that a choice has to be made between installation of equipment at the downstream and distribution of equipment all along the Seine.

Evidently, there is **no question of not utilising the capacity of transfer and storage of the networks.** In the case of utilising transfer capacity maximally, the capacity of treatment during the rainy season will have to be increased beyond 22 m³/s in the first section. On the other hand, it is difficult to say at the moment which solution is the most economical and whether the capacity of treatment during the rainy period at the Achères plant should be limited to only 22 m³/s or an additional section should be provided.

Utilisation of present equipment and planning future investment must be given serious thought. In the next 15 to 20 years it would be preferable to plan an equitable distribution along the Seine so that everyone can observe improvement.

If a different distribution is planned for combating run-off pollution, **it must be ensured that the tangible advantages accrued are commensurate with the financial obligation.** Aesthetics deserves particular attention.

In conclusion, Mr. **Affholder** expressed agreement that a solution must be found which meets the aspirations of the inhabitants, resolves requirements of the fish population and is globally economical.

As a first step, a tentative directive plan of sewerage was formulated in 1992. Improvements can be incorporated in this plan but one must await the results of the 'study of discharge' currently in progress.

Mr. **Mouchel** said that in view of the fact that everyone agrees that the Seine has to be kept clean and that the management of a network should take into account the effect of discharge on the Seine, it would be desirable that the management have access to **data collected by measurement stations** located all along the Seine. He asked Mr. **Le Scoarnec** whether the integration of these measurement networks with the system **'Partenaire'** is part of the short-term objectives.

Mr. **Le Scoarnec** responded that access to this system is relatively easy as 16 persons (partners) can use it simultaneously. **The only constraint is that the measurement network must be concentrated in one storage centre.** For example, it would definitely become possible to share with various partners the data relating to 9 points of measurement of dissolved oxygen in the Seine. In fact, all the data of these stations is stored at Colombes, which thus constitutes a data storage centre.

Mr. **Affholder** concluded that exchange of information with the APES network of Hauts de Seine is likewise possible. At present, however, the wherewithal for action against overflow of the Seine is very limited; only when such wherewithal and the requirements of storage and treatment capacities are met will management be able to integrate the state of the Seine in real time. Unfortunately, that stage has yet to be realised but it is envisaged in the next millennium.

Mr. **Gousailles** remarked that this system of exchange resembles a supermarket which stores a great deal of data from which anyone can choose whatever he likes. The difficulty is knowing the **value of the data.**

There are two possibilities: one, to make the entire stored data available after its validation and the second, to make the data immediately accessible as and when it is acquired by a central system. In the latter case the problem of quality and reliability of the data would arise.

Mme Cogez felt that **validity of data** could be assumed on the basis of the large amount of work carried out by SIAAP and Seine St Denis during the last two years. At stake is the release of only such data as can be profitably utilised; exchange of unusable data should be prohibited. The system 'Partenaire' should thus be able to determine the degree of quality, a position untenable until a few years ago.

Agreement to exchange data is the first step towards co-operation among partners. They must learn to work together. What they used to do during the dry period has made the 'network redundant' so to speak; they planned the 'works in real time' two years in advance. Much more effort and tenacity is required to do the same during the rainy season, but the gains would be considerable.

The day when one can manoeuvre from one value to another on the strength of its certitude will be a hallmark.

For the work planned for the months and years ahead, much more importance than given now will have to be attached to **operational costs** because budgeting problems will become more serious, as observed by Mme Cogez at Seine St Denis.

Mr. **Soulier** said that as for exchange of data, **a programme of homogenisation and preparation of a fixed format of exchange** (format SANDRE) has been launched in collaboration with the District of Hauts de Seine.

Mr. **Affholder** concluded the discussion on the exchange of data by remarking that these ideas have already been accepted and the work begun.

Mr. **Constant** sought the following information:

— What is the time limit for **bringing weirs into conformity** with the stipulations of the bylaw of 22 December, 1994?

— Will the bylaw of 22 December 1994 replace article 12 of the law on water?

— Lastly, are the two texts complementary?

Mr. **Verjus** clarified that the time limit for the bylaw of December 1994 is the same as that stipulated for the local communities to conform with the European Directive: until 1998 in the sensitive zone and 2005 for small agglomerations..... . These time limits are consistent with article 12 of the bylaw of 29 March 1993.

After thanking the audience for the interest shown in the discussion, Mr. **Affholder** invited Mr. **Tabuchi** to say a few words. Mr. Tabuchi concluded the meeting with the following remarks.

With respect to weirs, from the technical point of view there has never been a serious problem; solutions could always be found.

The major difficulty, as a matter of fact, concerns overall optimisation of the sewerage system.

That is why I suggested to Mr. Valiron that he title this volume 'Design and Management of Combined Sewerage Networks' rathen than 'Synthesis of Knowledge of Storm Weirs'.

This leads to a new management of the sewerage network in which measurements occupy a place of enhanced importance. One is on relatively certain ground when speaking of a network or, more specifically, hydraulics. But when it comes to the natural environment, the main problem presently is lack of a project manager; **the urgent need of the moment is to appoint a project manager for the natural environment.**

This brings me to another point: I recall a get-together about two years ago in Grenoble with J. M. Mouchel, C. Cogez, M. Gousailles and myself. We imagined a sci-fi scenario in which automated management was subordinate to the natural environment. During a storm event of 27 June 1999 while fish were dying en masse in the Seine, the president of SIAAP could sleep soundly were such a system in place.

Today there is SCORE and each district of the Parisian agglomeration has its own automated system of management. **So, it is puzzling why the problems are still far from being resolved.**

References

1. GENERAL WORKS AND DOCUMENTS

A. **Valiron et Tabuchi,** Maîtrise de la pollution urbaine par temps de pluie, Etat de l'Art, Tec. et Doc., Lavoisier, 1992.

B. **DEA Seine St Denis,** Guide des vannes, 1993.

C. **AGHTM,** Les stations de pompage d'eau, 4ème édition, Lavoisier, Tec. et Doc., 1991.

D. **Ministère de la Recherche,** Prise en compte des points singuliers i dans la simulation des écoulements dans les réseaux d'assainissement, janvier 1986.

E. **SHF,** La pluie source de vie, choc de pollution, Colloque SHF, 17-18 mars 1993, Houille Blanche n° 1/.2, 1994.

F. **Valiron,** Gestion des eaux, alimentation, assainissement, Presses de l'Ecole Nationale des Ponts et Chaussées, octobre 1989.

G. **Azzout, Barroud, Crès, Alfakih** (INSA), Techniques alternatives en Assainissement pluvial, Tec. Doc. Lavoisier, octobre 1994.

H. **Lyonnaise des Eaux, Valiron,** Mémento du gestionnaire de l'alimentation en eau et de l'assainissement, Tec. Doc, Lavoisier, septembre 1994.

I. **Fouquet-Coste, ENPC,** Evacuation des eaux pluviales urbaines, ouvrage collectif, 1978.

J. **Dupuy, Knaebel,** Assainir la ville hier et aujourd'hui, Dunod, 1982.

K. **Stahre - Urbonas,** Stormwater detention—Prentice Hall Inc., New Jersey, 1990.

L. **WRC,** Urban Pollution Management, 1994.

M. **AGHTM,** Spécial Eaux pluviales, Techniques-Sciences et méthodes (TSM), n° 11, novembre 1995.

2. REGULATIONS

a. **Loi sur l'eau,** n° 92-3, JO du 4 janvier 1992.

b. **Le Traitement des eaux résiduaires urbaines,** Directive de la Communauté Européenne, n° 91-271, JO de la CE L/35/40 du 30 mai 1991.

c. **Nomenclature des opérations soumises à autorisation ou à déclaration** en application de l'article 10 de la Loi n° 92.3 du 3 janvier 1993 sur l'eau, Décrets 93.742 et 93.743 (JO du 30 mars 1993).

d. **Circulaire CG 1333,** Instruction technique relative à l'assainissement des agglomérations, 22 février 1949.

e. **Instruction technique** relative aux réseaux d'assainissement des agglomérations, Circulaire interministérielle 77 284, 22 juin 1977.

f. **Décret 94.469** sur la réglementation spécifique à l'assainissement, 3 juin 1994.

g **Département de la Seine St Denis,** Règlement de l'assainissement départemental.

h. **Département de la Seine St Denis,** Règlement départemental de sécurité sur les réseaux d'assainissement.

i. **Arrêté du 22 décembre 1994** sur la fixation des prescriptions techniques relatives aux ouvrages de collecte et de traitement des eaux usées.

j. **Guide de recommandations** pour l'application du décret du 3 juin 1994 et des arrêtés du 22 décembre 1994 (circulaire du 12 mai 1995).

3. RECENT ARTICLES AND BOOKS

1. **Desbordes,** Bilan des études et recherches sur la pollution du ruissellement pluvial urbain dans les pays de l'Europe de l'Ouest et de l'Amérique du Nord, LHM, 1985.

2. **SIAAP - Score,** Bilan 1993, SGRA, 1994.

3. **Chebbo G. et al.,** Les bassins d'orage et la lutte contre la pollution des eaux pluviales, Journée d'étude des eaux pluviales, Agen, 1991.

4. **Bachoc, Chebbo G.,** Gestion des solides en réseau d'assainissement, décembre 1994, Groupe pluvial AGHTM.

5. **Bertrand-Krajewski J.L., Chebbo G.,** LED Cergrène, Le premier flot et le dimensionnement des ouvrages de traitement des rejets par temps de pluie, décembre 1994, Groupe pluvial AGHTM.

6. **Saget et Chebbo G.,** Analyse de la répartition des charges polluantes des rejets urbains par temps de pluie, Cergrène, novembre 1994.

7. **Geiger W.F.,** Flashing effects in combined sewer runoff, 4ème Conférence sur les eaux d'orage, p. 40-46, Lausanne, janvier 87.

8. **Cottet,** La pollution des eaux pluviales en zone urbaine, ENPC/ABSN, 1980.

9. **Tabuchi J.P.,** AESN, Les flux de pollution rejetés par les déversoirs d'orage, les gains possibles SHF GRAIE, Lyon, 17 novembre 1994.

10. **Hodeau D.,** Communauté Urbaine de Lyon. L'expérience de la Communauté Urbaine de Lyon, SAF GRAIE, Lyon, 17 novembre 1994.

11. **Ellis J.B.,** Urban discharges and receiving water quality impact, Pergamon press; 1990.

12. **Dowing and Merkens,** The influence of temperature on survival of several species of fish in low tensions of dissolved oxygen, An App. Bio 45, p. 231-267, 1957.

13. **Saul et Delo,** Laboratory tests on a storm sewer overflow chamber with unsteady flow, IAHR/IAWQ, 1982.

14. **Chocat B.,** INSA Lyon, Les différents types de déversoirs d'orage, Analyse et modélisation de leur fonctionnement, SHF GRAIE, Lyon, 17 novembre 1994.

15. **Mitri et James,** Program OVRFLO3, Open channel overflow discusion structures with side weit. Users manual, February 1982; CHI publication.

16. **LROP,** Etude du fonctionnement hydraulique de la Bièvre à Paris et de ses affluents. Déversoirs de Bièvre, Blanqui et Buffon, SIAAP, février 1989.

17. **Grange D., Ruban G.,** LROP, Ernploi d'une métrologie optique pour la mesure en continu de la qualité des eaux résiduaires urbaines. Journée des Sciences de l'ingénieur, LCPC, 47 octobre 1994.

18. **Delattre J.M.,** DEA Seine St Denis, La Gestion dynamique des déversoirs d'orage, SHF GRAIE, Lyon, 17 novembre 1994.

19. **Green J.,** UK experience with pollutions separation at combined sewer overflows, December 1993, NVRC Report 2226:

20. **Riding E.,** Monitoring of screens at combined sewer overtlows interim report n° l, School of construction, Sheffield Hulham University, July 1993.

21. **Walsh A.M.,** The Gross Solids Sampler. Development of a sampler for combined sewer overflows, WRC Report n° UM 1172, Nov. 1990.

22. **Jeffries C.,** Methods of estimating the discharge of gross solids from combined sewer systems, Wat. Sci., Tech. vol. 26, 1992.

23. **Walsch and Jeffries,** Final report on the evaluation of combined sewer overflows in Dunfermline (Scotland). Using the WRC Gross solids sampler, WRC report UM 1320, March 1992.

24. **Cootes T.F, Balmforth D.J., Graen M.J.,** Monitoring a vortex combined sewer overflow with peripheral spill WRC report UM 1169, Dec. 1990.

25. **Ruff S.J, Saul A.J, Walsh M., Green M.J.,** Laboratory studies of CSO performance, WRL report n° UM 1421, December 1993.

26. **Hedges P.D., Lockley P.E. and Martin J.R.,** The relationship between field and model studies of an hydrodynamic separator combined sewer overflow, IAHR/IAWQ Niagara Falls, Canada, pp. 1537-1542, 1993.

27. **Hedges P.D.,** "Combined sewer overflow devices". Seminar on combined sewer overflows, IHE Delft, Netherlands, 1994.

28. **Pisano et Brombach,** Demonstration of advanced high rate treatment for CSO control in metropolitan Toronto, Graie Lyon, p. 331-340, 1993.

29. **Marchandise P, Ruban G., Cathelain M., Wacheux H.,** LCPC AEAP Anjou, Recherche, Efficacité du traitement et du stockage d'eaux pluviales en réseau : évaluation par la mesure en continu, Lyon, Pollutec 92.

30. **Ellis J.B.,** Measures for the control and treatment of urban runoff quality, November 1991, Middlesex Pollution Research Centre.

31. **Crabtree R.,** WRC UK, Les préconisations anglaises pour le dimensionnement des ouvrages, SAF GRAIE, Lyon, 17 novembre 1994.

32. **Bellefleur D.,** ENGEE Strasbourg, Les innovations en matière de déversoirs d'orage, SAF GRAIE, Lyon, 17 novembre 1994.

33. **Agg and Derrick,** Storm sewage discharges to coastal waters in Netherlands, West Germany and France, Foundation for water research, Allen House, Dec. 1990.

34. **Voorhoeve,** Reduction de la pollution pluviale amenée au milieu naturel par les réseaux séparatifs et unitaires. Le cas des PaysBas, AESN, novembre 1991.

35. **NWRW,** Final report of the 1982-1988 research programme. Conclusion and recommendations, juin 1991.

36. **De Carne,** Note sur la situation en RFA, Colloque sur la pollution des eaux pluviales, AESN, Paris 1992.

37. **ATV 128,** Richtlinien für die Bemerssung und Gestaltung von Regenentlasungsanlagen in Mischwasserkanälen, Regelwerk, April 1990.

38. **FNDAE,** Ministère de l'Agriculture, AEAP, Les bassins d'orage sur les réseaux d'assainissement, Expériences requises à partir des réalisations actuelles, Mars 1988.

39. **EPA,** Methodology for analysis of detention basins for control of urban runoff, 1986.

40. **Safège,** Mike 11, Logiciel de modélisation des écoulements libres, Note explicative, 1989.

41. **Safège,** Mouse Logiciel de modélisation des réseaux d'assainissement, Note explicative, 1989.

42. **Bertrand-Krajewski,** Modélisation des débits et du transport solide en réseau d'assainissement, ENITRS, Lyonnaise des Eaux, avril 1991.

43. **Safège,** Note sur les quantités de pollution déversées à Metz par temps de pluie, avril 1992.

44. **Dégardin,** Quantification des rejets urbains de temps de pluie en région parisienne. Estimation des volumes et des flux de pollution déversés au milieu naturel lors d'une pluie décennale. AESN 19 pages plus annexes, mai 1991.

45. **Bachoc, Chebbo, Bonnefous,** Caractérisation des solides transférés dans le bassin de Beguigneaux, Cergrène, octobre 1990.

46. **Hansen O.B., Jacobsen C., Harremos P.,** Stormwater loading of greater Copenhagen sewage treatment plants. Water Sci. technol. N° 5-6, p. 49-59, 1993.

47. **Breuil,** DEA Seine St Denis, Besoins nouveaux, projets innovants, colloque "les bassins nouvelle vague", St Denis, 1617 juin 1992.

48. **NRA,** General guidances note for preparatory work for AMPZ, version 2, December 1993.

49. **Le Trionnaire Y.,** Diren Ile de France, Déversoirs d'orage, Le nouveau contexte réglementaire, SHF GRAIE, Lyon, 17 novembre 1994.

50. **ATV 117,** Abwassertechnische Vereinigung Richstlinien für die Bemessung und Gestaltung von Regenentlastungsanlagen in Mischwarser Kanälen Regelwerk, April 1990.

51. **Rumsey, Clifford, Crabtree,** Le contrôle de la pollution urbaine d'origine pluviale. Le programme de recherche dans la gestion de la pollution urbaine au Royaume-Uni. La Houille blanche, septembre 1993.

52. **Dempsey,** Urban pollution management. Planning guide for the management of urban wastewater discharges during wet weather. Foundation for the research, 1994.

53. **Crabtree, Gent, Clifford,** Contrôle de la pollution due aux déversoirs des réseaux unitaires. Expériences pratiques d'une gestion intégrée au Royaume-Uni, Techniques Sc. Méthodes n° 5, mai 1994.

54. **Balmforth, Saul, Clifford,** Guide to the design of combined overflows structures. Foundation for water research, report FR 0488, 1994.

55. **Harremoes,** Stockastic model for estirnation of extrem pollution from urban runoff. Water ressources, vol. 22, n° 8, 1988.

56. **Bujon,** Description du modèle mathématique Kalplan, AESN, janvier 1992.

57. **Clegg, Eadon, Fiddes,** UK State of the art, Sewerage rehabilitation Water Science Tech. 21-1101-1112, 1989.

58. **Pister B.,** DEA Val de Marne, L'expérience du Val de Marne, SAF GRAIE, Lyon, 17 novembre 1994.

59. **Balmforth D.J. and Henderson R.J.,** A guide to the design of storm overflow structures, WRC report n° ER 304 EAP, 1988.

60. **Van Lingtelaar H., Ganzevles P.P.,** Dynamic design of sewer detention tanks, technol. n° 5-6, p. 133-143, 1993.

61. **Rees A., White K.,** Impact of combined sewer overflows on the water quality of an urban watercourse Regal Rivers n° 1-2, p. 83-94, 1993.

62. **Foundation Chambers,** Déversoirs dans réseau unitaire, Récents développements en Europe, Clevedon, 1993.

63. **Mignosa P., Proletti A.,** Pollution loads discharged from combined sewer overflows. Theoretical approach and long term numerical assessment. Noril Hyrol, n° 1, p. 27-47, 1993.

64. **AGHTM,** Groupe pluvial. Elaboration des circulaires d'accompagnement des unités, 9 janvier 1995.

65. **DEA Val de Marne,** Le Val de Marne et la station de Valenton. Conception Générale, 29 juin 1993.

66. **DEG,** Quantification des déversements sur les principaux déversoirs parisiens, Paris, décembre 1993.

67. **Doyen,** La Molette, dépollution et décantation pluie du 7 août 1989, décembre 1990.

68. **Doyen and autres,** Modèle de décantation en bassin, application sur le bassin de la Molette.

69. **Département de la Seine St Denis,** DEA, Proposition pour un schéma d'assainissement, septembre 1992.

70. **DEA Seine St Denis,** Exutoires en Seine : la Vieille Mer, 30 mai 1994.

71. **DEA Val de Marne,** Le système Valérie, 29 juin 1993.

72. **IAWPRC,** Real time control for urban drainage systems : The State of the art. Ed. W. Schilhing, Pergamon Presse, Oxford, 1989.

73. **Green,** Benefits of real time control for urban drainage systems, WRC Report n° UC 1045, June 1991.

74. **Cant,** Detention tank design and maintenance, WRC Report UM 1233, 1991.

75. **Fahrner, Gresa,** Weitergehende Regenwasser behandlung mit Siebrechen an Entlastungsbauwerken, der Mischkanalisation.

76. **District de l'agglomération nancéienne, AERM,** Avant-projet de dépollution des eaux pluviales. Nancy - Hydratec, 1995.

77. **SIAAP,** Les barrages flottants, Bilan 1994 - SGRA - 1995.

78. **Hydratec,** Moka-Eau, Notice de programme, version 3.1., Seine, mai 1994.

79. **District d'Arras - AEAP - Anjou Recherche,** Schéma directeur de l'assainissement du district urbain d'Arras, Rapport de synthèse, juillet 1991.

80. **Serre,** Contribution à l'analyse des traitements des rejets de temps de pluie, Magistère Ingénierie et Gestion de l'environnement, ENSMP, ENGREF, ENPC.

81. **Paitry,** Le séparateur statique tourbillonnaire, Courbe hauteur débit et rendement, DEA Seine Saint-Denis, AFBSN-STU, 1984.

82. **Deneuvy,** La prise en compte des rejets urbains par temps de pluie dans la réglementation française dans TSM numéro spécial, novembre 1995.

83. **Bujon G.,** 1983. Modélisation de la dispersion de substances solubles ou pseudo-solubles dans un cours d'eau. Application au cas de la Seine. La Houille Blanche, n° 1.

84. **Chocat B., Cathelain M., Mares A. et Mouchel J.M.,** 1994. La pollution due aux rejets urbains par temps de pluie : impact sur les milieux récepteurs. La Houille Blanche 1/2 : 97-105.

85. **Edeline F.,** 1979. Epuration biologique des eaux résiduaires urbaines, théorie et technologie. Technique et Documentation, Paris. 306 pages.

86. **Estèbe A., Mouchel J.M. et Thévenot D.R.,** 1992. Impact des orages en zone urbaine sur les concentrations métalliques des matières en suspension de la Seine. In PIREN-Seine, rapport de synthèse 1989-1992, vol IV, "Impact des surverses d'orage sur la qualité des eaux de la Seine dans l'agglomération parisienne".

87. **Estèbe A., Bussy A.L., Videau V., Mouchel J.M., et Thévenot D.,** 1993. Teneurs en métaux des matières en suspension de la Seine. Rapport 1993/III du PIREN-Seine. CNRS.

88. **Even S. et Poulin M.,** 1993. Le modèle PROSE : hydrodynamique, transport et qualité de l'eau. In Actes du colloque "La Seine et son bassin : de la recherche à la gestion", CNRS, PIREN-Seine, 29-30 avril 1993, Paris.

89. **Even S., Poulin M., Mouchel J.M. et Billen G.,** 1993. Modélisation de la qualité des eaux de la Seine en période d'orage. Rapport 1993/III, CNRS, PIREN-Seine.

90. **Fisher H.B., List E.J., Koh R.C.Y., Imberger J. and Brooks N.H.,** 1979. Mixing in inland and coastal waters. Academic Press, 483 pages.

91. **Gammeter S. and Frutiger A.,** 1990. Short-term toxicity of NH_3 and low oxygen to benthic macroinvertebrates of running waters and conclusions for wet weather water pollution control measures. Water Science and Technology, 22 : 291-296.

92. **Garnier J., Billen G., Hanset P., Testard P et Coste M.,** 1993. Développement algal et eutrophisation dans le réseau hydrographie de la Seine. In Actes du colloque "La Seine et son bassin : de la recherche à la gestion", CNRS, PIREN-Seine, 29-30 avril 1993, Paris.

93. **Garric J., Migeon B. et Vindimian E.,** 1990. Lethal effects of draining on brown trout. A predictive model based on field and laboratory studies. Water Research, 24 : 59-65.

94. **Huang W.W. et Mouchel J.M.,** 1994. Concentrations totales et formes chimiques des métaux dissous dans la Seine. Rapport PIREN-Seine 1994/III. Groupe Bassins Versants Urbains.

95. **Mouchel J.M.,** 1993. Impact des rejets urbains de temps de pluie de l'agglomération parisienne sur les concentrations d' ammonium et d'oxygène dissous à Chatou, in Actes du Colloque "La Seine et son bassin: de la recherche à la gestion", CNRS, PIREN-Seine, 29-30 avril 1993, Paris.

96. **Servais P., Garnier J., Bariilier A. et Billen G.,** 1993. Dégradation de la matière organique à l'aval d'Achères, in Actes du Colloque "La Seine et son bassin, de la recherche a la gestion", PIREN-Seine, CNRS, Paris, 29-30, avril 1993.

97. **Simon L.,** 1995. Contribution to the modelling of surface water flow and transport. Thèse de Doctorat, Ecole Nationale des Ponts et Chaussées.

98. **Tangerino C.,** 1994. Mesure de la vitesse de chute des matieres en suspension, du carbone organique total et des bactéries en Seine. Rapport de DEA-STE, Université Paris-Val de Marne, 52 pages plus annexes.

Index